# CROSSRAIL

How the **Elizabeth line** was built

## Dan Harvey

Grosvenor House
Publishing Limited

The right of Dan Harvey to be identified as the author of this
work has been asserted in accordance with Section 78
of the Copyright, Designs and Patents Act 1988

This book is published by
Grosvenor House Publishing Ltd
Link House
140 The Broadway, Tolworth, Surrey, KT6 7HT.
www.grosvenorhousepublishing.co.uk

A CIP record for this book
is available from the British Library

ISBN 978-1-80381-881-8

# ABOUT THE AUTHOR

Dan Harvey reported on the Crossrail programme from funding approval in 2008 to completion in 2023. With a monthly column devoted to the project in *Modern Railways* magazine and regular updates on the Transport Briefing website he gained an unrivalled insight into the development of the Elizabeth line, one of Britain's largest ever rail infrastructure undertakings.

# CONTENTS

# PREFACE

I have a hazy memory of visiting the Museum of London as a child in the early 1980s and viewing a map of proposals for London's future public transport system. It was there I first encountered the name 'Crossrail' and the purple – my favourite colour – that has come to be associated with the new railway.

The idea of a new underground railway caught my interest but prompted bemusement; I saw London's transport system as a preserved relic, not something that could change. As a child only the Jubilee line stations in central London showed signs of recent investment. In my mind the London Underground had been handed down to us by a distant generation of railway builders and I assumed it would remain largely in the same form. I did not realise then that other cities in Europe were aggressively expanding and investing in transit. There was nothing to show that London's public transport system would change any time soon.

As a young adult this view persisted. I spent a lot of my time commuting – from Hemel Hempstead and later Croydon. Rail privatisation brought new logos, and much needed new trains – but routes and infrastructure changed little. There was, however, the intriguing story of Thameslink – British Rail had somehow managed to reinstate a short section of disused line that allowed tracks south and north of the river to be connected, creating a north-south, through London train service. Not only could you travel through London, without having to pick up connecting Tube trains and schlep between stations, but you could travel direct from the suburbs to central London stations like Farringdon (for Smithfield, Hatton Garden), City (St Pauls, Fleet Street) and Blackfriars.

Thameslink was ludicrously successful. Despite being a relatively new service peak travel was uncomfortable and very, very overcrowded. In the evening peaks commuters crammed on board at the London stations so that on departure north, from the now defunct King's Cross Thameslink, and south from London Bridge, trains were rammed.

When I landed a publishing job in Croydon my initial plan was to commute cross-London from St Albans. However, this commuting experience proved so stressful that I soon relocated south of the river.

It was clear that major investment was needed and plans for Thameslink 2000, later the Thameslink Programme, and Crossrail edged forwards. I was interested and started following the detailed development of the plans. Soon I was sharing updates through an email newsletter and the Transport Briefing website. When Crossrail got the go-ahead there was suddenly lots happening on the transport investment front. I felt the information available online was hard to find and poorly packaged so with Transport Briefing and spin-off website Crossrailnews I had a go at filleting and contextualising the press releases to show how the big rail investment projects were moving forwards. *Modern Railways* magazine commissioned me to write a monthly Crossrail column (thanks James) which I would file every month until the new railway was completed.

As Crossrail activity gathered pace the initial, rather limited official website was replaced with an excellent and comprehensive online presence providing information throughout the duration of the Crossrail programme. Yet, now that it has been completed, most of this has been archived and appears not to be readily available.

This then is my story of how the Elizabeth line was built, gleaned from unpicking the announcements, visiting construction sites and talking to key people involved in the programme. It's an exciting story of building something vast out of nothing; bequeathing something to future generations – just like the London Underground pioneers. It's an absorbing story of detail – when you realise how clever the components of Crossrail are it's difficult not to get sucked in. It's a story to learn from – how people work or don't work together, how to spend (or not spend), and how technology allows us to unlock the potential of a nineteenth century innovation while making it harder to deliver then ever before.

Crossrail has proved that it is possible to build a showcase railway through the middle of one of the most well know cities on the planet. I fear, however, that after spending £20 billion it will discourage others from embarking on a similar approach. Key questions – What makes a successful project leader? How can taxpayer funded projects be held to account? – need answering to encourage confidence in future ventures.

Perhaps the biggest issue to be resolved is that of technology. Crossrail applied a modern overlay to a technology forged in the industrial revolution. But despite this challenge being identified early on it emerged that few people understood the technology and even those who did didn't necessarily know how to make it work. This display of technical ignorance when it comes to the humble railway system is fascinating when set beside the current rapid developments in artificial intelligence. We don't all need to know how things work but we need a reliable cadre of experts who do.

This book attempts to set out clearly and honestly how the Elizabeth line was built. Not everything went to plan and not everything deserves a positive spin. However, travel on the Elizabeth line today, through the new tunnels, the vast new stations, and you will see for yourself that Crossrail is an astonishing achievement.

## A note on the text

Thank you to all those who have contributed information or insights that are included in this book. Please forgive any omissions, inconsistencies or inaccuracies which are all my own.

For updates on the text and Elizabeth line developments visit crossrailbook.co.uk.

# 1. THE MAKING

That exhibition at the Museum of London all those years ago provided a glimpse of the railway to come. Not that it would happen any time soon; despite Crossrail being seen as integral to modernising London's rail network it was achingly slow to get off the starting blocks.

Crossrail's roots can be traced back to planning during the Second World War for how Britain should look once hostilities ceased. The Abercrombie plan, which resulted in the development of New Towns and safeguarding green belt, included proposals for two new railways across London.

These were not built but as the London economy expanded the need for improved transport links became more pressing and the ideas of the post-war planners were revisited. The 1974 London Rail Study, published by the then Greater London Council and Department for Environment, included the word 'Crossrail' – neatly summing up the ambition of building a railway across the capital.

Although the envisaged new central London Underground interchanges at Paddington, Bond Street and Liverpool Street were not dissimilar to those found on today's Elizabeth line, as was the proposed connection of existing overground lines to the west and east of the capital, plans for a tunnel connecting the Southern Region's central services at Victoria, Blackfriars and London Bridge were quite different to the scheme eventually built.

Indeed, a wholesale return to the drawing board would be required but, in a defining moment for Crossrail, the study said no to the idea of building a new London Underground line. Compare and contrast the Victoria line, completed in 1972, with the Elizabeth line: it's not just the scale of platforms and stations which is so different. Crossrail would be a metro railway with full size trains – and that meant the new underground railway could join up with existing overground lines.

Not only did that open up new journey opportunities to and through central London but it also cut out a major logistical block for the railway by doing away with the need to turn around trains – and build stations with several platforms to accommodate these movements. When it came to Crossrail London would not be at the end of the line.

This is an idea which is now embedded in rail planning across Europe and back in the 1970s the concept was gaining traction. The Réseau Express Régional (RER) in Paris and the German S-Bahn systems in Hamburg and Munich were now having a bearing on transport planning for London. In the 1980s the creation of Thameslink with the reinstatement of a short section of railway linking Farringdon and Blackfriars achieved the first cross-London metro railway but the Elizabeth line is the first time the new-build, full-size, joined-up railway concept has been realised in London.

Grand plans are a start but who would pay? With Crossrail estimated to cost £300 million the answer was – in the 1970s – no-one.[1] But in 1980 a British Rail discussion paper proposed an inter-city link across London featuring three route options and costed at £330 million. Crossrail was back.

At this point the scheme lacked clear direction both in terms of project organisation and geographical alignment with Crossrail rotated ninety degrees to link existing rail infrastructure north and south of London. Route, however, was of secondary importance to the ratification of the 1974 proposal's aim to deliver a full size railway that would allow trains from the suburbs to run underground through London.

In a repeat performance at first these proposals too looked set to languish. But after decades where a lack of investment in the Underground system had come to be referred to as 'managed decline' there was a sense that money would need to be spent. The King's Cross fire in 1987, where 31 people died, was linked to dated

---

[1] This princely sum would later prove insufficient to complete even the planning phase of Crossrail

wooden escalators plus lax workplace and public safety practices. Modernisation – and a cash injection – was long overdue.

With the number of journeys made on the London Underground increasing year on year it became increasingly apparent that existing Tube and rail capacity was insufficient to meet the demand for travel. This led to the government commissioning and then publishing the Central London Rail Study in 1989 – a report which repackaged many of the schemes considered by the 1974 study to present a menu of rail interventions that could be used to respond to rapid growth in commuter traffic. In his introduction to the report Transport Secretary Paul Channon acknowledged that this was putting severe strains on London's transport system. Multiple Crossrail options were mentioned in the report including what eventually became the Elizabeth line and a line from Wimbledon to Hackney via Chelsea (now safeguarded as Crossrail 2).

## Conservatives back CrossRail

The Conservative Party conference, in October 1990, saw the government instruct British Rail and London Underground to work up plans for the east-west Crossrail scheme. With an estimated price tag of £1.4 billion work was projected to start in 1993 so that – prizes for positive thinking – the new cross-London line could open in 1999. BR and LU resurrected and refined previous plans for the project and, in November 1991, a private members bill for CrossRail with a capital R was ready to begin its parliamentary journey.

With Network South East branding being rolled out across the railways radiating from the capital CrossRail quite logically appropriated the corporate red and blue colours on its maps and mock-ups of Class 341 Networker rolling stock.[2] Despite the difference in liveries compared to today's trains a look at those maps shows a familiar route from Shenfield in the east and on through the central London Elizabeth line stations used by passengers today to Maidenhead and then Reading.

---

[2] This was exhibited to commuters at Liverpool Street station

Reading? The Department for Transport and Transport for London only committed to running Crossrail trains to Reading in March 2014. Yet back in the 1990s that was all part of the CrossRail plan.

But something big was missing from the route. In the early 1980s Conservative minister Michael Heseltine had been a driving force calling for the regeneration of the derelict Docklands area and setting up the London Docklands Development Corporation. Between 1981 and 1988, encouraged by tax incentives for businesses, more than £8 billion of private money was invested in the area. This saw Docklands emerge as an economic hub and wealth generator for London and beyond. Public transport would in turn be needed to support the new homes and businesses and in 1987 the Docklands Light Railway opened for this purpose.

But CrossRail made no provision for London's emerging Docklands district or the Canary Wharf estate that would be built on the Isle of Dogs. Even considering the drawn out development of the scheme it seems, with hindsight, that transport planners were slow to grasp the opportunities presented by Docklands regeneration.

Without Docklands the map of 1990s Crossrail was a mirror image of the Elizabeth line today. No Canary Wharf or Abbey Wood meant Shenfield was the only eastern terminus. To the west the new railway would run not only to Reading and Heathrow but also Aylesbury and Chesham using part of the routes served today by Chiltern Railways.[3]

Whatever the ideal route of a new railway it was not going to happen without serious funding. After welcoming the conclusions of the 1989 study Paul Channon was supportive of Crossrail but he cautioned more work was needed to identify where the money for such a scheme could come from. Along with revenue from future ticket sales he suggested property developers who would benefit from the new infrastructure should contribute towards the cost.

---

[3] Chiltern's first franchise agreement, signed in 2000, allowed for the train operator to be involved in the development of Crossrail

This sounded reasonable enough – except when a slowdown in the economy struck, reducing demand for rail travel and making developers less able or willing to find cash for transport improvements. The recession also constrained public finances and gave the Treasury one more reason not to commit to a major outgoing that would last for years.

Estimated CrossRail costs had reached £2 billion by 1993. Cost increases on the recently approved Jubilee line extension project dented confidence that another big rail scheme could stay within budget and ministers looked for opportunities to cut costs. The scheme progressed notwithstanding this and, following his surprise General Election win in 1992, Prime Minister John Major championed the project even if the Treasury was less convinced.

## The end of the beginning

Despite this high-level support 11 May 1994 would prove a pivotal moment in Crossrail history. The private bill had been put before a House of Commons 'Opposed Bill' committee which, despite having only four members, would have the power to decide the fate of the scheme. This antiquated parliamentary mechanism, soon to be abolished, put key decisions in the hands of a small group of people – as well as those in government responsible for nominating members of the committee. Tony Marlow (Conservative), Ken Purchase (Labour) and Dr John Marek (Labour) all voted against the bill. Although Matthew Banks (Conservative) voted in favour this simple three against one vote was enough to sink hopes of the 1990s CrossRail programme being built.

Looking back it seems almost nonsensical that so few people were able to torpedo such a big, and potentially far-reaching scheme at an unremarkable panel meeting that happened to be scheduled as part of an arcane process. Yet that is what happened and hopes of Crossrail becoming part of British Rail and Network South East ended here.

That's not to say the decision was popular; more than 200 MPs signed a petition disagreeing with the committee's ruling. Many agreed that halting CrossRail, a scheme designed to underpin long term economic growth, was the wrong decision. This was shown when, even as costs increased, the government continued to support the scheme. Yet the window of opportunity had by now shut; John Major had supported CrossRail but his decision to privatise Britain's railways represented a major blow to the scheme. With so much railway reorganisation to think of infrastructure schemes became an unwelcome distraction. Railtrack, which was taking over British Rail's responsibilities in a major industry restructure, had neither the time nor the motivation to push CrossRail forward.

Who was to blame for the failure to get CrossRail off the ground? Major? Rail privatisation? Or that small group of MPs on the Opposed Bill committee who rejected the plans under an old-fashioned parliamentary instrument which, thankfully, no longer operates.

But perhaps the Opposed Bill Committee members had solid grounds for their decision. They spent several months considering 314 petitions from opponents to the scheme or people who wanted to see it modified in some way. Maybe the plans had not been sufficiently thought through. Certainly the preparation for 1990s Crossrail was more rudimentary than that for the scheme which eventually went ahead. Decision-making structures were informal. One member of the original leadership panel told me that at meetings it was like the mafia getting together.

Delving into the detail of the scheme one starts to question if the project was ready – or was it rushed? How strong were the economic and regenerative credentials of CrossRail without a Docklands station? Would new electrification be required on the Aylesbury route? Where was the promised private sector finance? Come to think of it where was the evidence of any sort of robust funding package?

## Picking up the pieces

If there was some consolation it was that progress had been made. Despite the decision not to go ahead the October 1990 announcement meant the Crossrail alignment was now safeguarded from property ventures that could thwart future construction. A strong case had been made to build a new east-west railway and members of parliament and business representatives sought ways to bring the scheme back to life.

Just two months after the Opposed Bill Committee rejected Crossrail, plans were unveiled by London Transport and British Rail to take forward the scheme using the Transport and Works Act, which had become law in 1992. This allowed the Secretary of State to "make an order relating to, or to matters ancillary to, the construction or operation of a transport system" including a railway. Where parliamentary process had proved intractable the Transport & Works Act offered a lifeline to keep Crossrail alive.

But it would not move forward quickly. Treasury inertia, linked to a desire to avoid committing funding, meant a new study was commissioned to see if there was a cheaper alternative to Crossrail offering similar benefits. The resulting study said there was not. At the same time the government reassessed the economic case for Crossrail and concluded that the private sector alone could not be relied on to come up with the money needed for the project.

As is often the case with rail infrastructure schemes in Britain these exercises squandered both time and money which perhaps could have been better spent building something. But it did at least prove that the underlying case for Crossrail was sound and sensible assumptions had been made about what the structure of the project could look like.

Crossrail development continued but in the 1990s you might have been hard pressed to notice. These days Transport & Works Act orders are not an everyday occurrence but the application procedure is well established for both light and heavy rail developments. Nearly two years after the rejection of CrossRail, and the decision to follow the TWA route, there was little sign of an application being lodged. Then in April 1996 Secretary of State for Transport Sir George Young asked London Transport and British Rail to halt work on their TWA application reasoning that Crossrail could not move forward until development of the Jubilee line extension,

Thameslink Programme and the Channel Tunnel Rail Link were further advanced.

With that the project would remain on the shelf for the rest of the decade. On the plus side, the east-west cross-London route had been officially safeguarded. A small team within London Underground continued to work on Crossrail; as planning applications were lodged for major property developments along the route these people made sure foundations for new office blocks would not get in the way of building Crossrail in the future.

## New Labour, new opportunity

Not for the first time the wider restructuring of Britain's national rail system, along with a change to the organisation of London transport, would be closely linked to the pace at which Crossrail could move forward. By 2000 a Labour government was in power and the Strategic Rail Authority had been created, outsourcing responsibility for rail network planning from Whitehall. There would later be tension between the SRA and ministers leading to government clawing back control but, at this point in time, the establishing of the SRA with its opportunity to reexamine rail strategy was good news for Crossrail.

The SRA was asked by government to look at the need for extra passenger capacity to and through London and the resulting London East West Study recommended that development of both Crossrail and the Hackney-Chelsea Crossrail 2 routes should be taken forward. While this was happening Ken Livingstone became the first Mayor of London through a devolution settlement that handed him significant transport powers; the first Mayor's Transport Strategy made the relief of overcrowding on the Underground a priority with Crossrail held up as the means by which to achieve this.

Finally, the forces needed to take forward Crossrail had aligned. But despite a consensus on the need to increase capacity on the London and south east rail network it took eight years for the government and parliament to be persuaded that Crossrail would be a cost-effective means of addressing this need.[4] Still, action was forthcoming. Cross London Rail Links Ltd, a joint venture between government owned SRA and Mayor of London Ken Livingstone's Transport for London, was set up to define the requirements of an east-west Crossrail study and undertake a feasibility study of a possible Crossrail 2 scheme. With a budget of £154 million Cross London Rail Links became operational in January 2002.

By July 2003 the new organisation had submitted a Crossrail Business Case to Transport Secretary Alistair Darling. The Secretary of State signalled the government's support for the principle of a new east-west rail link across the capital but commissioned Adrian Montague to look into the financial viability of such a scheme. The Montague report, published in July 2004, said that raising the estimated £10 billion cost needed for the programme would be a huge challenge. Undeterred, the following day Alistair Darling said the government would introduce a bill "at the earliest opportunity" to obtain the powers needed to construct Crossrail.

By this time the route of Crossrail's central London core was largely accepted. What was still undecided was how this would connect up to the existing rail network to create the Paris RER-style through London metro which was so fundamental to the Crossrail concept. CLRL's stakeholder briefing document of 2002 shows that to the west of London Crossrail lines to Watford Junction and Aylesbury, as well as Reading and Heathrow, were live options. A branch serving Docklands was by now part of the Crossrail business case but the thinking was that trains would join one of the existing Kent lines before Plumstead and then run on to Ebbsfleet, connecting with the new Ebbsfleet Channel Tunnel Rail Link station.[5]

In 2003 CLRL was asked by the Transport Secretary to launch a public consultation on the Crossrail proposals when the options presented included a scheme serving Heathrow and Kingston in the west and Shenfield and Ebbsfleet in the east, connected by new

---

[4] First National Audit Office Crossrail report: Crossrail, 24 January 2014
[5] CTRL is now known as High Speed 1

tunnels under central London. Between August and October 2004 a second consultation updated proposals with a service to Maidenhead but dropping the Kingston branch. CLRL cut the south eastern branch back to Abbey Wood in November 2004 because of fears that Crossrail trains sharing track with north Kent services further east could have a negative impact on reliability.

Labour's 2004 spending review passed without a government commitment to fund Crossrail but there was a promise to put together legislation that would facilitate construction of the new railway. The government also pledged to work with TfL, and other stakeholders eager to see the scheme go ahead, to develop a funding package. At CLRL Keith Berryman, on-off chief executive in the early 2000s,[6] and Douglas Oakervee, executive chairman between December 2005 and May 2009, oversaw the drafting of Crossrail legislation and handled the painstaking revisions and refinements that were required before the project was fit to be considered by lawmakers in Westminster. They also worked to acquire the land required by the programme.

## The journey through parliament

Wheels had been set in motion. In February 2005 a Crossrail bill was deposited in parliament and, following the false start of the private members bill, the decision was made to opt for a hybrid bill.[7] Although not used often hybrid bills are suited to large rail projects; before Crossrail a hybrid bill was chosen to secure powers to construct the Channel Tunnel Rail Link and the parliamentary instrument has subsequently been used with High Speed 2.

A key selling point for scheme promoters is that a hybrid bill allows flexibility as proposals are refined. As the bill proceeds through its

---

[6] Keith Berryman was acting chief executive of CLRL at launch and held a number of senior roles on the project including deputy chief executive. He was appointed acting chief executive again in August 2005. Crossrail appoints acting chief executive, Building, 10 August 2005

[7] The Transport and Works Act was later used to acquire powers to supplement the provision of what became the Crossrail Act

stages in parliament the government has scope to make revisions although the question as to whether or not the scheme is a good idea in principle is established at an early stage. This prevents opponents of a project using examination of details to question the merit of the overall enterprise. A further benefit is that hybrid bills make it more difficult to trigger a judicial review, a delaying tactic often used by those opposing infrastructure projects.

Crossrail's second parliamentary journey, as hybrid bill, bore little resemblance to the experience of the private members bill more than a decade earlier. Process meant it would be considered by both the House of Commons and House of Lords, each of which would have three readings (each one getting successively shorter), with opportunities to back, reject or amend the bill. A select committee stage considered petitions – concerns raised by members of the public – and made recommendations for changes where appropriate. Douglas Oakervee would be called upon throughout the process to provide clarification whenever a query emerged concerning an element of the programme design.

In July 2008, three and a half years after being deposited in parliament, the hybrid bill received Royal Assent. This put the Crossrail Act 2008 on to the statute book – effectively providing the planning powers needed to construct the long-awaited new railway across the capital.

Lord Bassam, one of the two MPs selected to promote the Crossrail bill in parliament, paid tribute to all those people who had shaped "what is probably one of the most exciting engineering and rail projects that this country has on the stocks and in the immediate and near future – a project that will very much influence how London continues to generate and regenerate itself in the next two decades".[8]

---

[8] Lord Bassam, Third Reading of Crossrail Bill in the House of Lords, prior to Royal Assent

### How the Crossrail Bill became law

"Following the introduction of the Crossrail Bill in the other place [the House of Commons] in February 2005, some three and a half years ago, the Bill received a Second Reading by a majority of 370. After a period of gentle persuasion, the Select Committee in the other place was appointed in December 2005 and first met in January 2006. There were four batches of additional provisions to improve the project, which is one reason why the committee took nearly two years to conclude its hearings. It heard in excess of 200 of the 400 or so petitions lodged; its special report was published on 23 October 2007.

"The Public Bill Committee, Report and Third Reading in the other place followed expeditiously and the Bill was introduced into this House on 14 December 2007. The Second Reading was secured without a Division and, by the time the petitioning period for this House's Select Committee had closed, 113 petitions had been deposited. The Select Committee of this House did a sterling job; it sat for 29 days and listened to many of the same petitioners."

Lord Bassam of Brighton

Perhaps the most significant change made to the Crossrail bill on its journey to becoming law was the addition of Woolwich station on the south east Elizabeth line branch, between Canary Wharf and Abbey Wood. The select committee considering the Crossrail bill agreed (after lobbying from Greenwich council and the local member of parliament) that a station at this location would not only provide solid evidence of Crossrail regenerating communities through which it would pass but that it could also offer value for money. Anxious to contain project costs the government was reluctant to embrace the plan. However, Transport Secretary Douglas Alexander announced in March 2007 that, subject to property developer Berkeley Homes making a financial contribution, Woolwich station would be added to the Crossrail bill.

This inclusion aside, the Crossrail signed off by MPs remained based on the east-west plans put forward decades before. The Crossrail

Act 2008 confirmed Crossrail as a railway between Maidenhead (not, at this stage, Reading) and Heathrow in the west to Shenfield and Abbey Wood in the east featuring new railway tunnels and stations in the West End and Docklands.

Looking back over an extraordinarily protracted 34 years of planning and development it's remarkable how little Crossrail changed. But finally, no longer just an aspiration, Crossrail had made the jump to fully fleshed out project signed off by the law-makers of the land. Main construction work would start less than a year later.

At long last Crossrail had approval and could be built. But before the builders could begin one not-so-small loose end remained. How would Europe's largest rail project of the time be paid for?

Connaught Tunnel approach. The Crossrail Act made provision
for the old North London line route to be restored
for use by Elizabeth line trains.

# 2. THE MONEY

The first thing to grasp about the cost of Crossrail is that generally, when referred to, it does not include the Elizabeth line trains nor the construction and maintenance of the main train depot at Old Oak Common in west London. For those you need to add on an extra billion and a bit pounds; when the train and depot contract was signed in February 2014 the value quoted by Bombardier was £1.3 billion. Note, however, that with financing and train maintenance costs factored into the contract the line between 'up-front' and operational costs is blurred.

With that out of the way we can concentrate on the numbers relating to the Crossrail programme infrastructure – the materials and labour that have been required to build a new underground railway and upgrade existing surface lines to create a coherent 'Elizabeth line'.

During the long gestation of Crossrail there was plenty of opportunity to discuss where the money should come from. Among the points made – that people and businesses in London who would benefit from the project should contribute to the cost; that the private sector should pay towards the cost of the new railway; and that ultimately – for a project on the scale of Crossrail – the government would have to underwrite the venture.

With the Crossrail hybrid bill making its way through parliament funding was finally nailed down in the heads of terms agreement of November 2007. This presented Crossrail as having an estimated capital cost, including contingency, of £15.9 billion. This is sometimes referred to as 'the funding envelope' – how much money was available for the project rather than how much it would actually cost. Responsibility for finding this sum was split largely between the Department for Transport and the Mayor of London. Since early development of the scheme Crossrail was jointly sponsored by the DfT and the Mayor and the funding model mirrored this.

The sources of money in the heads of terms agreement can be seen over the page. In summary, the DfT would put in a little over £5 billion which, taken with third party contributions, would see it take charge of finding £5.6 billion. The Mayor, as the leader of London government, would oversee input of £7.7 billion coming from various income streams available to the London Assembly and mayoral agency Transport for London. The greater part of this was expected to come from the National Non-Domestic Rate business supplement with a further slug from TfL – pre-Covid-19 pandemic largely funded from ticket revenue – plus taxes on new property developments. Crossrail would be the impetus for new and updated mechanisms for collecting this cash.

Together DfT and TfL cash promised £13.3 billion for Crossrail. The remainder would be sourced from Network Rail and would be used to pay for 'on-network works' – the upgrades needed to the existing overground rail network to ensure the Crossrail route could operate. Network Rail, which under European accounting rules has since 2013 been deemed to be a public sector organisation, receives track access fees paid by train operators (and funded from ticket receipts or government subsidy) plus money direct from the government.

As one would expect – big project, delivery over many years – Crossrail costs have changed and the project's sponsors have had to respond to any changes in circumstances and fill any funding gaps that emerged. In early 2010, with construction of the new railway at an early stage, the estimated capital cost had soared to £17.8 billion. Following the election of the David Cameron-led coalition government in May 2010 this provided a prolonged period of uncertainty for Crossrail – would the new government recommit to Crossrail? For many months trade journalists pored over every word from ministers that might signal a green or red light for the project and wrote up their latest Crossrail news story accordingly.

Eventually – in October – the coalition said yes to Crossrail. It had taken the opportunity to put on display the supposed financial discipline of the new administration – there was much talk of 'rigorous cost control', 'no sacred cows' and 'value engineering'. In the first half of 2010 Crossrail Ltd, the TfL subsidiary charged with delivering the programme, identified savings of £1.4 billion that it said could be achieved by reducing risk before main construction began and by re-sequencing the programme – the order of digging tunnels and stations. In June 2010, as part of a comprehensive spending review, the Treasury demanded more savings from the Department for Transport which then worked with Crossrail to take a further £1.6 billion out of the programme. This would be achieved by:

- Reducing risks by simplifying integration works, re-sequencing work and removing scope, saving £800 million; for example, deciding not to create a direct connection from the Elizabeth line to the District and Circle Line platforms of the London Underground at Paddington station.
- Lowering the costs of inflation to reflect the impact of recession upon economic forecasts, saving £300 million.
- Agreeing contracts with lower target prices, as a result of the global recession which encouraged construction companies to deliver very competitive bids, saving £500 million.[1]

In total £3 billion was stripped off the £17.8 billion Crossrail programme cost quoted in 2010. Not only did this bring the project back within the £15.9 billion funding envelope agreed in the 2007 heads of terms but it provided an opportunity, rightly or wrongly, for the government to demonstrate financial stewardship by reducing the cost to £14.8 billion. In addition to this updated funding envelope there was a promise at the time to deliver Crossrail for no more than £14.5 billion. Not only was the programme going ahead but, thanks to this administration, it would actually be built for less!

The £14.8 billion funding envelope would remain intact for the next eight years and meant, at least on paper, the DfT and Transport for London would be off the hook for about half a billion pounds each. Still, they had their work cut out to find the cash; how would it be collected?

---

[1] First National Audit Office Crossrail report: Crossrail, 24 January 2014

# Breakdown of funding sources in 2007 heads of term agreement

| | £bn |
|---|---|
| **TFL UNDERWRITTEN** | |
| GLA (national non-domestic rates debt) | 3.5 |
| TfL core contribution | 2.7 |
| London Underground interface savings | 0.4 |
| Sales of surplus land and property | 0.5 |
| Developer contributions | 0.3 |
| London Planning Charge | 0.3 |
| **DFT UNDERWRITTEN** | |
| DfT grant contribution | 5.1 |
| BAA / City Corporation (guaranteed) | 0.5 |
| **OTHER** | |
| Network Rail (on-network works) | 2.3 |
| Depot (operating lease) | 0.5 |
| City Corporation (additional) | 0.1 |
| Less other residual costs | (0.4) |
| **TOTAL SOURCES (equal to capital cost including contingency)** | 15.9 |

Notes:
1. Sources and uses breakdown captures items classed as capital only
2. Contribution from Canary Wharf Group included within assumed purchase price for Isle of Dogs station
3. Figures rounded to one decimal place
4. All figures expressed in nominal terms
5. TfL developer contributions includes contributions from section 106, Planning Charge and Wood Wharf
6. £300 million Planning Charge contribution underwritten by Her Majesty's Government

## DfT

The Department for Transport was responsible for leveraging, loaning and underwriting many parts of the Crossrail funding package but the £5.1 billion was a fairly straightforward, bog-standard, vanilla government grant, the single largest component of Crossrail funding.

During the passage of the Crossrail bill through parliament the government made a commitment to publish an annual update showing how much had been spent on Crossrail during the past 12 months and cumulatively. While the data conflates DfT and TfL spending on the project, and also periodically includes loans to Network Rail for work on Crossrail, we can see below that over a decade typically £1 billion to £1.5 billion was spent a year on the programme by the joint sponsors.

**Annual expenditure (£) by Crossrail Ltd[2]**

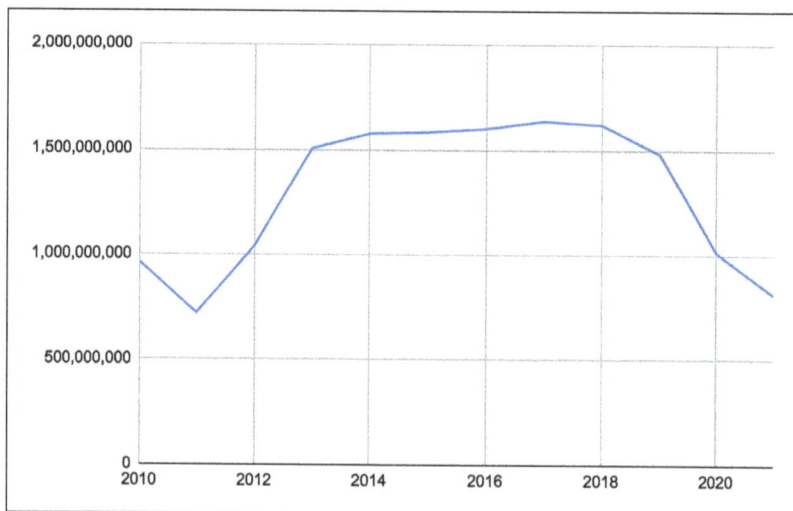

---

[2] Data from DfT annual Crossrail statements. See table above.

18

## Crossrail annual statements published by the Department for Transport[3]

| Period covered by statement | Total funding amounts provided by Crossrail Ltd by the DfT and TfL in relation to the construction of Crossrail (from 22/7/08) | Expenditure incurred including committed land and property spend not yet paid out) by Crossrail Ltd in relation to the construction of Crossrail during this period | Total expenditure incurred (including committed land and property spend not yet paid out) by Crossrail Ltd in relation to the end of the period (from 22/7/08) | Amounts realised by the disposal of any land or property for the purposes of the construction of Crossrail by the Secretary of State, TfL or Crossrail Ltd during the period |
|---|---|---|---|---|
| 2010 | £773,769,000 | £963,976,000 | £1,160,779,000 | £0 |
| 2011 | £1,484,605,000 | £723,475,000 | £1,884,254,000 | £0 |
| 2012 | £2,751,624,360 | £1,044,294,000 | £2,928,548,000 | £0 |
| 2013 | £4,258,541,482 | £1,506,347,000 | £4,434,895,000 | £0 |
| 2014 | £5,981,006,309 | £1,576,835,000 | £6,011,730,000 | £0 |
| 2015 | £7,952,017,859 | £1,583,293,000 | £7,650,559,000 | £0 |
| 2016 | £10,002,888,670 | £1,599,948,000 | £9,250,507,000 | £0 |
| 2017 | £10,860,539,046 | £1,636,471,000 | £10,886,978,000 | £0 |
| 2018 | £11,713,723,131 | £1,619,238,000 | £12,506,215,837 | £18,462,238 |
| 2019 | £13,165,913,790 | £1,481,243,170 | £13,958,459,007 | £143,778,674 |
| 2020 | £14,164,813,354 | £1,014,218,000 | £14,972,678,000 | £16,000,000 |
| 2021 | £14,893,427,506 | £813,125,000 | £15,785,802,000 | £0 |
| Total | £14,893,427,506 | £15,562,463,170 | £15,785,802,000 | £178,240,912 |

The disclosures to parliament also include amounts realised by the disposal of any land or property during the Crossrail progamme. The 2007 heads of terms envisaged half a billion pounds being raised this way.

In addition to the core DfT grant contribution the department pledged to negotiate with the City of London Corporation and Heathrow Airport to secure contributions from businesses towards

---

[3] Period refers to the year ending 29 May

Crossrail. Whatever the outcome of these negotiations the heads of terms committed the DfT to providing half a billion pounds on top of the £5.1 billion already pledged.

The City of London Corporation initially offered to make a contribution of £200 million, a figure that subsequently increased to £250 million.

At the same time as the Department secured the guaranteed £250 million funding from the City of London Corporation, the Department for Communities and Local Government agreed to reinstate an arrangement whereby it allowed the City of London to retain around £10 million a year from its contribution to the national business rates pool. The reinstatement of this arrangement took into account the planned Crossrail contribution by the City; over time this additional funding will offset the City's contribution to Crossrail.[4]

In addition the City of London said it would seek voluntary contributions from businesses in the capital with the aim of raising a further £150 million. Note that this was on top of the DfT's committed funding. According to Crossrail Ltd the City of London Corporation voluntary funding agreed is £100 million. This is accounted for in the heads of terms underneath the 'other' subheading.

Less successful have been attempts to get Heathrow Airport – which now has three stations served by the Elizabeth line – to put money into the project. The heads of terms refers to BAA – the company that owned several UK airports, including Heathrow, prior to being directed by the Competition Commission in 2009 to sell off the other airports in its portfolio.[5]

---

[4] First National Audit Office Crossrail report: Crossrail, 24 January 2014
[5] After divesting the other airports the BAA name was replaced with Heathrow Airport Holdings. In 2024 HAH was owned by a consortium made up of infrastructure firm Ferrovial (25%), Qatar Investment Authority (20%), Caisse de dépôt et placement du Québec (12.62%), Singaporean sovereign wealth fund GIC (11.2%), Australian Retirement Trust (11.18%),

In 2008 the Department expected Heathrow to pay £230 million, based on an estimate of the benefits Crossrail would bring to the airport. This contribution was subject to the approval of the Civil Aviation Authority, the airports' regulator. Heathrow later decided that, with Heathrow Airport operating at or near to capacity, Crossrail would bring no net benefit to the airport, but made a provisional allowance of £100 million in its final business plan. In summer 2013 the Department made a counterproposal of £137 million. In January 2014 the CAA determined that Heathrow should contribute £70 million, less than a third of the amount envisaged.[6]

## Mayor/Transport for London

Although the DfT grant was the single largest component of the Crossrail funding package, the Mayors of London – and there have been three of them during the development of the programme – have overseen a larger contribution to the making of the Elizabeth line. When the heads of terms were drawn up it was envisaged that the Mayor's TfL would underwrite funding totalling £7.7 billion.

Included in this was a core contribution from Transport for London. As operator of much of London's public transport system TfL would be able to use part of its investment budget set aside from fare revenues (although the Covid pandemic showed that this revenue could not be taken for granted). The heads of terms also loosened the restrictions around TfL borrowing so long as the organisation could afford to service the debt.

In 2007 it also foresaw savings being found where Crossrail interfaces with London Underground infrastructure – as is the case at several stations. The heads of terms envisaged a £100 million saving in 2013 with further savings of £150 million in each of 2014 and 2015. Much has changed in TfL's financial position since then with the government requiring the organisation to find many more

China Investment Corporation (10%) and Universities Superannuation Scheme (10%). See https://www.heathrow.com/company/about-heathrow
[6] First National Audit Office Crossrail report: Crossrail, 24 January 2014

savings. As can be seen in chapter 10, these LU interface savings are no longer itemised in the funding breakdown.

The largest component of Crossrail funding falling under the Mayoral/TfL umbrella was the national non-domestic rate business supplement. Introduced in April 2010, the business rate supplement is charged at 2p per £1 on commercial properties with a rateable value over £55,000 in the Greater London Authority area. It is collected on the same bills as business rates and, as is the case for rates, is payable by a range of organisations with premises including the NHS, police, schools and TfL. Prior to the relocation of the GLA to The Crystal at the end of 2021, the GLA was expected to contribute almost £100,000 through the BRS each year to Southwark Council for use of City Hall.

This is a long term funding mechanism and it is expected that the supplement will be collected over a 30 year period which is not expected to end until 2038. The Greater London Authority has been required to borrow up front to meet Crossrail costs and the Business Rate Supplement will be used to pay this back with interest. The thinking behind the BRS is that, as a form of local taxation, the business rate income is less sensitive to economic circumstances than individual private sector contributions. Seventy per cent of BRS contributions were expected to be paid by ratepayers in London boroughs which have a station on the Crossrail route.

According to the heads of terms £3.5 billion is predicated on the business rate supplement. This is the amount that the GLA will make available through borrowing and from 2019-20 BRS receipts were expected to be used to pay back the loan. In addition £600 million of the BRS income in the first five to six years was due to be paid directly to TfL to finance the construction costs. This equates to the expected amount by which the BRS income in the first few years exceeded the interest costs on the GLA's £3.5 billion of borrowing.[7] Together, this adds up to £4.1 billion, referred to as

---

[7] Crossrail Business Rates Supplement Q&A, Greater London Authority, 2010

business rate supplement, borrowing and direct London contribution on Crossrail's updated funding breakdown (see chapter 10).

## Developer contributions

Also overseen by the Mayor is the £600 million stated in the heads of terms which was due to be found from a combination of developer contributions and a London planning charge, two similar sources, both forecast to bring in £300 million, which are easy to confuse. The developer contributions relate to a long-established process of property developers agreeing to pay local authorities a sum in recognition of the local infrastructure (including transport) or other amenities that is required to underpin a major development. These are often referred to as section 106 contributions.

Whereas these developer contributions are open to a certain amount of negotiation, the London planning charge was an attempt to implement a more regimented payment system – a tax – on commercial property developments such as office space, retail and hotels that benefit from investment in new local infrastructure.

When the heads of terms was announced the government agreed to underwrite the new statutory planning charge with the Mayor of London to implement an appropriate payment system. The result was the Community Infrastructure Levy, introduced in London in 2012 and since referred to as the Mayoral CIL, distinguishing it as a Crossrail specific tax, separate from other CILs which have since been introduced. Local authorities in London collect the MCIL, which is calculated according to the floorspace of a development, and passed on to the Mayor.

Since its introduction the CIL has proven to be a reliable way of collecting money for new infrastructure and the model has been adopted by local authorities outside London. A revamped charging schedule, introduced in 2019, allows money collected through the MCIL to be set aside for Crossrail 2.

During construction of Crossrail these two different planning charges have had mixed fortunes. Section 106 contributions have been slower to materialise than expected and underpinned by a contribution for the Wood Wharf development, to the east of Canary Wharf and more recently known as 'Canary Wharf's New District'. On the other hand Mayoral Community Infrastructure Levy payments have exceeded expectations; by August 2017 nearly £400 million had been collected for Crossrail, exceeding the amount set out in the heads of terms.[8] This meant that when the Elizabeth line opened planning contributions well in excess of £600 million had been collected for Crossrail – even if the lion's share had been sourced from the MCIL.

## Network Rail

Funding for the Elizabeth line's overground sections has been a roller coaster ride. Numbers went up, down, up again and what was included in the price of the ticket has regularly proved subject to change.

The 2007 heads of terms gave Network Rail a budget of £2.3 billion to deliver on-network works. In September 2009, however, Network Rail told Crossrail Ltd that it would need £3.1 billion to do this work.[9] Over the next year Network Rail brought its costs to within the available £2.3 billion after further detailed design and work to identify efficiency savings. This included reducing £200 million payable in compensation to train operators through more effective line possessions and avoiding 'gold-plating' at stations. Plans to upgrade Ilford and Romford stations were scaled back.

In 2013 NR said it had brought the cost down to £2.15 billion but had agreed to take on extra work providing power supplies for the central tunnelled sections via overground sub-stations. This would bring the cost of NR back up to £2.3 billion.[10]

---

[8] Crossrail update, Modern Railways, August 2017
[9] First National Audit Office Crossrail report: Crossrail, 24 January 2014
[10] Crossrail Supplement, Modern Railways, 2013

Alongside working out the most affordable way to deliver the Crossrail on-network works Network Rail has, as custodian of stations, been regularly intertwined with initiatives to deliver additional improvements to the Crossrail network. This has included ensuring all overground Elizabeth line stations offer step free access to platforms – funding for this was awarded in 2014 – and extra station works, commissioned by TfL to rectify station buildings and structures which had fallen into disrepair under Network Rail and precursor organisations Railtrack and British Rail.

Crossrail is primarily a publicly funded undertaking with the money needed to build it largely coming from central and local government, which in turn are funded by taxpayers. However, there has been some private sector cash input. Heathrow Airport, as previously mentioned, has committed £70 million. Additional money has been secured to deliver two of the new Crossrail stations.

## Canary Wharf

A major new Elizabeth line station – but not built by Crossrail? Canary Wharf station is in many ways unique.

Property developer Canary Wharf Group had long been an advocate for Crossrail and once the project moved towards approval it argued, successfully, that with 20 years experience delivering more than 15 million square foot of office space, it was perfectly positioned to deliver the new station.

So in December 2008 the Department for Transport agreed to outsource construction of what, until Crossrail agreed to CWG's request for a name change on 14 May 2009, was known as Isle of Dogs station. Canary Wharf Group's construction arm, Canary Wharf Contractors, would start building the station that year.

The arrangement extended, to a certain extent, to financing the station project. Canary Wharf Group would receive £350 million from the Crossrail funding package with the developer contributing a further £150 million, although an agreement would allow it to offset this against any future Crossrail section 106 contributions for certain agreed Docklands development sites. Furthermore, Canary Wharf Group would bear the risk on the project should the station cost more than the estimated £500.3 million forecast.

We don't know exactly what Canary Wharf Elizabeth line station ended up costing to build; one consequence of private sector involvement is that it is easier to invoke claims of commercial confidentiality to keep details private. The arrangement initially appeared to have brought in £150 million of private sector cash for Crossrail. However, this was conditional on the railway opening within a designated time window and in 2020 Transport for London finance boss Simon Kilonback told the London Assembly Transport Committee that a confidential agreement had been reached to take account of subsequent delays.

However the finances pan out, CWG didn't hang around: Canary Wharf station was the first Crossrail station to start and finish main construction. However, as discussed in chapter 9, subsequent problems meant it was one of the last stations to be handed over to the infrastructure manager ready for passenger service.

## Woolwich

Woolwich station has been delivered by Crossrail and its contractors but as a late addition to the programme it was only going to happen if private sector money was forthcoming.

During the passage of the Crossrail bill through parliament the Labour government of the day resisted calls to specify the extra station, worried that it could add £300 million to the cost of the new railway. Ultimately campaigners successfully argued that omitting a station from an area in need of regeneration would be an opportunity lost when the railway was going to be built through Woolwich anyway. The station was added to the Crossrail Bill in 2007.

However, the Labour government only agreed to this on condition that Berkeley Homes, responsible for a major house building and regeneration scheme at Royal Arsenal Riverside, would bear the cost and risk of constructing the station box. Berkeley (pronounced 'Barkley') and other stakeholders would also have to fund the station fit out.

A final funding agreement to secure construction of the box was signed by Berkeley and the Crossrail sponsors in February 2011. As with Canary Wharf the specifics of the deal and precisely how much Berkeley committed have not been disclosed but the Woolwich station box was expected to cost around £100 million.

This left the question of who would pay for station fit-out unresolved. The issue came to a head in 2013 with Crossrail Ltd warning that it needed a decision by July to ensure that the station could be fitted out cost effectively and in time for Crossrail opening and test train running. Postponing fit-out risked importing the extra costs and disruption of working on an operational railway.

In August 2013 a funding deal to fit out Woolwich was agreed: Crossrail Ltd would be paid a fixed sum of £54 million consisting of £10 million from Berkeley Homes, £20 million from the Royal Borough of Greenwich and £24 million from Transport for London.

Greenwich noted that council tax payers would not have to contribute to the fit-out costs with £15 million of its contribution to be sourced from property developers through the Community Infrastructure Levy and the remaining £5 million coming from a Greater London Authority grant. Indeed, by the time fit-out of Woolwich station had been completed Greenwich had yet to pay most of the amount it had predicated on CIL receipts.

Crossrail Ltd estimated the total cost of Woolwich station fit-out at between £75 million and £100 million but this included some work at the North Woolwich and Plumstead portals. It said it would meet the difference between this and the £54 million funding package out of money set aside for ventilation and emergency exit works which would be required at Woolwich with or without the station.

As we have seen the finances underpinning Crossrail are complex; the scale of the scheme was too big for a single grant or handout to carry it forward. Instead, discrete funding elements have been welded together – each one painstakingly negotiated, prodded and subject to revisitation at any point to see if this money could still be counted on. But where the promoters of the previous CrossRail scheme failed, this time round a workable package was agreed.

The story of the money does not end there. Projects evolve, costs change and promoters must adapt – so it was for Crossrail and chapter 10 picks up this financial thread. But in 2008, when the Crossrail Act became law, funding to allow the Elizabeth line to be built was in place. Planning powers – check. Money – check. Time now to get on with building the new railway.

Canary Wharf station structure assembled.
The Crossrail funding package saw construction and
financing outsourced to Canary Wharf Group.

# 3. THE COMPANIES

Who would – who could – build Crossrail? As the funding package fell into place Cross London Rail Links, and subsequently Crossrail Ltd, would need to overcome the challenges of mobilising an extensive supply chain. The scale of work was such that contracts needed to be phased to avoid overwhelming the number of suppliers available and exceeding market capacity. Specialised, technical requirements would also constrain the client organisation's room for manoeuvre; for example, if you want to buy a tunnel boring machine there are not that many options.

For companies, the ramping up of a vast supply chain requiring everything from mobile offices to platform screen doors offered a gold rush of opportunity. Big construction firms were presented with an almost unprecedented selection of contracts to bid for and across London, wherever Elizabeth line activity was scheduled, myriad sub-contracts would be available to bid for through online portals such as CompeteFor.

Throughout Crossrail procurement and construction Crossrail Ltd was eager to stress the programme's procurement credentials and maintain support for the project. Yes, London would benefit from the build but companies across the UK would win work on the Elizabeth line; from trolleys for tunnel boring machines being built in Newcastle Upon Tyne to the purple roundels for stations being made on the Isle of Wight.

Thousands of contracts were let for the Elizabeth line. Crossrail said it procured 116 major contracts with a combined value of more than £8 billion.[1] For each of these 'tier one' contracts work would be sub contracted to create a multi-layer supply chain with packages available for companies of all sizes. As well being able to source big ticket items, the tier one contractors would

---

[1] Crossrail Learning Legacy, Procurement

procure day to day construction industry essentials such as aggregate supply and cleaning services.

But it is the tier one contracts which are of particular interest here, especially those pertaining to construction of the new railway. For me this is partly because I'm intrigued to know where does one go if in the market for 81 new escalators but also because, given the challenges we saw in chapter 2 of identifying the billions needed to build Crossrail, I feel we should follow the money. Who did it go to? Even better if we could ascertain if this was money well spent.

Thirty-six main contracts (numbered in bold in the following tables), with a value of at least £6.5 billion, were awarded to a handful of companies working either alone or in joint ventures.[2] These included, in no particular order, BAM Nuttall, Ferrovial Agroman, Kier, Dragados, John Sisk, Hochtief, Murphy, Balfour Beatty, BeMo Tunnelling, Morgan Sindall, Siemens, Vinci, Costain, Skanska, Alstom, VolkerFitzpatrick, Knorr Bremse, Laing O'Rourke and TSO. These companies brought expertise, innovation and dedication to delivering the Elizabeth line. But they also invoiced for significant chunks of the programme budget and claimed more money when delivery became challenging. Innovators and experts certainly but all these companies exist to make a profit. Crossrail's challenge would be to chart a mutually beneficial path that would allow companies to make a reasonable return while not fleecing the citizens that, one way or another, were paying for the new railway.

---

[2] Crossrail project: commercial aspects of works contracts for London's Elizabeth line, ICE, November 2017

# Key Crossrail tier one contractors[3]

| Who? | What? | Shares? |
| --- | --- | --- |
| Alstom | French multinational which supplies rolling stock and railway systems including signalling. In 2021 it completed the purchase of Canadian firm Bombardier's Transportation division. | Paris listed |
| Balfour Beatty | British construction and infrastructure company which as well as building provides financial and asset management services. | London listed |
| BAM Nuttall | UK construction and civil engineering company which is owned by Royal BAM Group, the largest construction firm in the Netherlands. | Amsterdam listed |
| BeMo Tunnelling | Austrian tunnelling, geotechnical and construction company. | Privately owned |
| Bombardier | Canadian company which won the contract to build Crossrail trains at its factory in Derby but now operates predominantly in the aviation sector after selling its transportation arm to Alstom. | Toronto listed |
| Carillion | Former British construction and facilities management company which collapsed in 2018 after major financial problems within the firm emerged. | No longer trading |
| Costain | British construction and engineering company founded in 1865. | London listed |
| Dragados | Construction and infrastructure business owned by Spanish conglomerate ACS Group. | Madrid listed |
| Ferrovial | Spanish multinational with portfolio of infrastructure interests including airports and a construction arm, known as Ferrovial Agroman before being rebadged Ferrovial Construction. | Madrid listed |
| Hochtief | German construction and concessions company. Main shareholder is Spain's ACS Group. | Frankfurt listed |

(Table continued)

---

[3] Company details correct at time of publication but are subject to change

(Table continued)

| Who? | What? | Shares? |
|---|---|---|
| Kier | British construction, services and property group with experience of financing projects. | London listed |
| Knorr Bremse | German braking specialist which supplies a range of rail systems. | Frankfurt listed |
| Laing O'Rourke | Formed from a merger in 2001, now Britain's largest privately owned construction business with a range of subsidiary companies geared to supporting project delivery. | Privately owned |
| Morgan Sindall | British construction, infrastructure and property company. | London listed |
| Murphy | Established by John Murphy in 1951 to work on post-war regeneration schemes, now a multi-disciplinary construction firm operating in the UK, Ireland and Canada. | Privately owned |
| Siemens | German multinational conglomerate, one of the biggest companies trading on the Frankfurt stock exchange. Supplies a vast range of products and technologies including trains and railway systems such as signalling and electrification. | Frankfurt listed |
| Sisk Group | Sisk is a private construction and property company founded by John Sisk in Cork, Ireland in 1859. | Privately owned |
| Skanska | Swedish multinational construction and project development company. | Stockholm listed |
| Travaux Sud Ouest (TSO) | French rail infrastructure specialist, a key supplier to SNCF, owned by French construction group NGE. | Majority employee owned |
| Vinci | French construction and concessions group with interests spanning different infrastructure sectors. | Paris listed |
| VolkerFitzpatrick | British construction and engineering contractor, a subsidiary of Dutch company VolkerWessels, owned by the Wessels Family through Reggeborgh Holding. | Privately owned |

## Programme and project delivery partners

Once Crossrail received Royal Assent Cross London Rail Links – and from 2009 Crossrail Ltd[4] – faced the daunting prospect of creating a vast project delivery organisation from next to nothing. The scale of the railway enterprise would need the skills of thousands of people, many with specialist skills. Identifying the best people, amid competition from other infrastructure schemes around the globe, would be challenging and aspirations to deliver 'world class' engineering demanded careful selection from within a finite human resources pool.

As well as finding the right people Crossrail needed to rapidly develop commercial expertise. Contract awards to major engineering companies drafted in much-needed building know-how but also experienced commercial teams who had been in position far longer than Crossrail's rapidly created client organisation.

To give Crossrail the commercial acumen it would need to deal with construction heavyweights a decision was taken early on to appoint project and programme delivery partners. These contracts would give Crossrail access to experienced, private sector staff from international project management organisations which would help the publicly set up project organisation handle the tier one construction contracts.

In April 2008 Cross London Rail Links advertised delivery partner work valued at between £375 million and £550 million. Lot 1, entitled Programme Delivery Partner, was for a company to help deliver the overall Crossrail project to time, budget and adhering to quality targets. Lot 2, described as Project Delivery Partner, was for an organisation to be responsible on a day-to-day basis for delivery of the central tunnel section of the Crossrail route. Lot 1 was valued at between £75 million and £150 million while Lot 2 was between £300 and £400 million.

---

[4] In November 2008 it was announced that Cross London Rail Links would change its name to Crossrail Ltd in the new year after securing the rights to the name from promoters of the Glasgow Crossrail scheme

In March 2009 Crossrail Ltd selected Transcend as its Programme Partner. Transcend was a joint venture between Aecom, CH2M Hill and Nichols Group and saw off bids from a Bechtel/Halcrow/Systra joint venture and Legacy 3, a Parsons Brinckerhoff, Balfour Beatty Management and Davis Langdon joint venture.

The Programme Partner contract, valued then at approximately £100 million, would see Transcend take on overall management of the Crossrail scheme on behalf of client Crossrail Ltd "with the specific purpose of acting as the programme partner helping Crossrail Ltd fulfil its obligations to deliver safely the overall programme to time, to the desired standard and within budget, as well as supplementing the core Crossrail Limited team with technical capability".[5] Nichols, which worked on the Crossrail programme in various capacities for more than 34 years, described its role as a strategic adviser and extension of the client's team. "We bring world class professionals with expertise in all aspects of programme management, working collaboratively with Crossrail providing leadership, strategic insight and delivery support across several workstreams including programme assurance, technical assurance, systems integration and addressing commissioning issues that were required to open the new railway."[6]

In April 2009 US engineering giant Bechtel was confirmed as Crossrail's Project Delivery Partner, supported by its nominated sub-suppliers Halcrow and Systra. Under this contract Bechtel was responsible for overseeing delivery of the core central tunnelled section of the Crossrail route and would also manage the design process. The San Francisco headquartered company at the time owned one third of Tube Lines, the consortium tasked with upgrading the Jubilee, Northern and Piccadilly Underground lines, and was part of the Rail Link Engineering consortium responsible for constructing the Channel Tunnel Rail Link, now High Speed 1.

---

[5] Crossrail Learning Legacy, Programme organisation and management
[6] Nichols Group website, July 2023

Three other bids had been shortlisted for the project delivery partner role: Legacy 3, a joint venture between Balfour Beatty/Parsons Brinckerhoff/Davis Langdon Programme Management, Laing O'Rourke Holdings and a joint venture between Capita Symonds and Northcroft.

For the first two years of their appointments both partners worked with Crossrail Ltd to establish the organisation needed to deliver Crossrail, including its management systems and control functions. Mobilising the large workforce needed to undertake these tasks was a major challenge.[7]

## Value added?

The project and programme delivery partner roles would be among the longest-running Crossrail contracts and account for significant expenditure. Much of that was on talent – to bring in the people with the skills needed to drive forward the project meant paying market rates. While Crossrail Ltd was required to publish details of its highest paid employees the contractors did not face such constraints.

Project and programme delivery partners are a common feature of big infrastructure programmes and there's a good reason for that. It is difficult to imagine Crossrail construction proceeding so quickly without drafting in external resources – expert organisations that can access the people and skills needed at a particular point in a project lifecycle. However, they come at a price and can create inefficient working practices and a lack of transparency.

During the early stages of Crossrail construction there were concerns of 'man-marking', staff employed by Crossrail Ltd and the delivery organisations effectively fulfilling the same role with confusion over who was responsible for what. As part of negotiations in the run-up

---

[7] Crossrail Learning Legacy, Programme organisation and management

to review point 4 in 2011,[8] a new structure was agreed which attempted to bring together Crossrail Ltd, project delivery partner Bechtel and programme partner Transcend as a seamless organisation which cost less to run. Crossrail Ltd said both organisations provided access to the important skills and experience needed to complete the programme.[9] The new structure would ensure there were no gaps or overlaps, and that each job was only done once.

However, having originally hired Bechtel and Transcend as partners to support it in managing the overall programme – including integrating the work of multiple contractors – folding the Bechtel and Transcend teams into its own project management effort, rather than holding them at arm's length, effectively stripped them of accountability for integration of the overall programme.[10] From this point on Bechtel and Transcend were largely invisible with staff embedded within a single 'Crossrail' organisation.

So long as the programme had access to the skills it needed and milestones were being met this change looked immaterial. But later on, when it became clear that delivery of the Elizabeth line was neither on time nor on budget, it seemed less appropriate for Bechtel and Transcend – which had been brought on board to make the programme run smoothly – to continue to be submitting invoices.

In 2019 the Department for Transport and Transport for London approved new incentive arrangements for Bechtel by repurposing £34 million of previously unearned incentives to help retain staff and to incentivise achieving the milestones set out in Crossrail Ltd's August 2020 delivery plan. In March 2021 the Chair of the Elizabeth line committee approved changes to Transcend's incentive

---

[8] The point at which Crossrail Ltd was authorised to award major contracts without needing to seek further permission from its sponsors – the Department for Transport and Transport for London
[9] Second National Audit Office Crossrail report: Completing Crossrail, 3 May 2019
[10] Second National Audit Office Crossrail report: Completing Crossrail, 3 May 2019

arrangements. Between the start of the contracts in 2009 and 31 March 2021 Crossrail Ltd paid £531 million to Bechtel and £127 million to Transcend.[11]

## Design contracts

In December 2008 Crossrail announced that 12 engineering consultants had secured a place on its design framework, enabling them to compete to win packages of design work on tunnels, shafts, stations, and railway systems.

The 12 successful companies were Aedas, Arup, Atkins, BDP, Capita Symonds, Halcrow, Hyder Consulting, Jacobs, Mott MacDonald, Parsons Brinckerhoff, Scott Wilson and WSP. Three teams – Faber Maunsell/Gifford, Mouchel and Scott Brownrigg – who were on the shortlist announced in August 2008, were not selected for the design framework (although Faber Maunsell successor Aecom was subsequently awarded design work on Wallasea jetty).

While a position on the framework was not a guarantee of work, most of the companies securing a place would go on to win Crossrail design contracts, as can be seen in the table over the page.

Crossrail opted for a consistent design for station platforms but, by appointing different station designers, signature designs for station entrances and ticket halls. This is an approach that worked well for the Jubilee line extension but may have added to Crossrail costs by specifying a wider range of fittings and bespoke building techniques than was required.

The winners of the station design contracts each teamed up with a firm of architects (details in chapter 5) to develop their specification for Crossrail. Design contracts were also awarded for components that would feature across the Crossrail programme as well as specific elements such as the trains for the new line and the tunnelling academy.

---

[11] Third National Audit Office Crossrail report: Crossrail – a progress update, 9 July 2021

## Design contract awards

| Contract number | Name | Awarded to |
| --- | --- | --- |
| 100 | Architectural components | Atkins |
| 102 | Material and workmanship | Mott MacDonald |
| 121 | Sprayed concrete lining | Mott MacDonald |
| 122 | Bored tunnels | Arup |
| 123 | Intermediate shafts | Jacobs |
| 124 | Railway systems and tunnel mechanical and electrical | Mott MacDonald |
| 125 | Mechanical and electrical in tunnels | Mott MacDonald |
| 130 | Paddington station | Scott Wilson |
| 131 | Paddington Integrated Project | Mott MacDonald |
| 132 | Bond Street station | WSP |
| 134 | Tottenham Court Road station | Arup |
| 136 | Farringdon station | Scott Wilson |
| 138 | Liverpool Street station | Mott MacDonald |
| 140 | Whitechapel station | Hyder |
| 146 | Custom House station | Atkins |
| 150 | Royal Oak Portal | Capita Symonds |
| 152 | Pudding Mill Lane Portal | Scott Wilson |
| 154 | Victoria Dock Portal | Hyder |
| 156 | North Woolwich and Plumstead Portals | Capita Symonds |
| 160 | Rolling stock and depots | Mott MacDonald |
| 161 | Ilford depot stabling | Mott MacDonald |
| 162 | Signalling, traction power, overhead line electrification and platform screen doors | Mott MacDonald |
| 164 | Bulk power distribution and high voltage power | Scott Wilson |
| 170 | Communications and control systems | Parsons Brinckerhoff |

(Table continued)

(Table continued)

| Contract number | Name | Awarded to |
|---|---|---|
| 175 | Tunnelling academy design | Capita Symonds |
| 176 | Wallasea temporary jetty | Aecom |
| 178 | Westbourne Park bus depot | Capita Symonds |

## Enabling contracts

In April 2009 Crossrail announced that 17 companies had secured a place on its enabling works framework, making them eligible for work packages designed to prepare for the main Crossrail contracts that would be awarded later. The framework was divided into four categories – site facilities, demolition, civil structures and utilities.

The companies appointed to the framework are listed below with the categories they were eligible to bid for contracts from shown in brackets.

- Balfour Beatty (Civils)
- BAM Nuttall (Site facilities, Demolition)
- Brown & Mason (Demolition)
- Carillion Civils Engineering (Civils)
- Clancy Docwra (Utilities)
- Costain/Skanska Construction joint venture (Civils, Utilities)
- Fitzpatrick Contractors (Site facilities)
- J Murphy and Sons (Civils, Utilities)
- John F Hunt Demolition (Demolition)
- Keltbray (Demolition)
- Kier Construction (Demolition, Civils)
- Laing O'Rourke Construction (Demolition, Civils, Utilities)
- McGee Group (Demolition)
- McNicholas (Utilities)
- Morgan Est (Civils, Utilities)
- PJ Carey (Demolition)
- Select Plant Hire (Site facilities)

The majority of these enabling works were carried out through 22 separate NEC3 Option A contracts (a priced contract with an activity schedule).[12] In addition to the enabling works contracts Crossrail let other early works contracts to specialist companies, outside the enabling works framework, to deal with hazardous materials, technical requirements and archaeological excavations. These, and the enabling works contracts, are listed below.

### Enabling and early work contract awards

| Contract number | Name | Awarded to |
| --- | --- | --- |
| 200 | Pudding Mill Lane site facilities | BAM Nuttall |
| 201 | Royal Oak Portal taxi facility – demolition | Keltbray |
| 203 | Pudding Mill Lane demolition | Laing O'Rourke |
| 207 | Bond Street demolition | McGee |
| 208 | Tottenham Court Road demolition | McGee |
| 209 | Farringdon east demolition | Keltbray |
| 210 | Liverpool Street advance works package 2 civil engineering | Murphy |
| 212 | Liverpool Street demolition | John F Hunt |
| 213 | Liverpool Street relocation of signalling | Signalling Installation Maintenance Services |
| 214 | Liverpool Street mechanical and electrical | RGB Integrated Services |
| 215 | Liverpool Street cable identification | RGB Integrated Services |
| 216 | Liverpool Street combined services works package 2 utilities | Laing O'Rourke |
| 217 | Whitechapel works package 2 civil engineering | Carillion |

(Table continued)

---

[12] Institution of Civil Engineers, Crossrail project: commercial aspects of works contracts for London's Elizabeth line, November 2017

(Table continued)

| Contract number | Name | Awarded to |
|---|---|---|
| 218 | Pudding Mill Lane hazardous materials surveys | WYG |
| 219 | East security/building maintenance site facilities | BAM Nuttall |
| 220 | Old Oak Common site facilities | Select Plant Hire |
| 221 | West security/building maintenance site facilities | Select Plant Hire |
| 223 | Bond Street east site facilities | Select Plant Hire |
| 225 | Liverpool Street site facilities | Select Plant Hire |
| 230 | Westbourne Park demolition | Morgan Sindall |
| 233 | North Woolwich and Victoria Dock CAW utilities | Murphy |
| 240 | Bond Street utilities | McNicholas |
| 243 | Liverpool Street combined services works package 1 utilities | Subsumed into C503 |
| 244 | Whitechapel works package 1 civil engineering East London line steel platform | Kier |
| 245 | Whitechapel civil engineering and utilities | Murphy |
| 248 | Pudding Mill Lane Portal civil engineering (including TBM reception chamber) | Costain/Skanska |
| 251 | Paddington Eastbourne Terrace utilities | Laing O'Rourke |
| 252 | Limmo Peninsula gas diversion utilities | McNicholas |
| 253 | Advance works hazardous materials surveys central section | RPS |
| 254 | Archaeology (combined west) | Oxford Archaeological Unit |
| 257 | Archaeology (combined central) | Museum of London Archaeology |
| 261 | Archaeology (combined early east) | Museum of London Archaeology |

(Table continued)

(Table continued)

| Contract number | Name | Awarded to |
|---|---|---|
| 262 | Archaeology (Pudding Mill Lane) | Wessex Archaeology |
| 263 | Archaeology (combined late east) | Museum of London Archaeology |
| 271 | Paddington Integrated Project – civil engineering | Carillion |
| 272 | Paddington Integrated Project – main works | Carillion |
| 277 | Old Oak Common demolition | Keltbray |
| 280 | Old Oak Common temporary bus depot | Balfour Beatty |
| 298 | Plumstead Depot | Balfour Beatty |
| 330 | Royal Oak Portal | Costain/Skanska |

## Contract structure

In 2008 a decision was made to let most of the major contracts to deliver the Elizabeth line using the NEC3: Engineering and Construction Contract Option C format – contracts featuring a target price with activity schedule.[13] This form of contract depends on contractors meeting deadlines to hand over work sites to other contractors and to minimise the volume of change and compensation events that require commercial agreement between parties. It also requires tight control and management by the client.

This type of contract was well known – it had been used to prepare for the 2012 Olympic Games and build High Speed 1, the

---

[13] NEC is a family of contracts, developed by the Institution of Civil Engineers, that are designed to support project management and define legal relationships between client and contractor.

42

format was endorsed by the government, TfL and the Institution of Civil Engineers (ICE) and the template, while robust, provided sufficient flexibility for different Crossrail contract packages.[14]

During construction of any major project circumstances dictate that costs will sometimes rise; conversely opportunities may present themselves to achieve savings. The NEC Option C contract recognised this by building in a pain/gain share mechanism. Crossrail opted for a 50/50 split as providing a fair balance between client and contractor with the advantage that it was easy to understand; any extra costs or savings achieved would be halved and paid by or refunded to both parties.

Some changes were made to the standard NEC3 Option 3 terms. These included requiring tier one contractors to use project bank accounts from which they would pay their sub-contractors. To discourage contractors from delaying payments to their suppliers any interest earned was paid to Crossrail.

Differences of opinion between client and contractor are inevitable with complex construction contracts. To resolve any disputes Crossrail set up an adjudication panel under the auspices of ICE. Crossrail encountered a number of disputes with contractors, mostly revolving around issues of time, but generally engaged in a process of managerial discussions to reach agreement.[15] As of November 2017 only one formal adjudication had been made in relation to the main works contracts.

Overall, the choice of NEC3 Option C benefited Crossrail by providing visibility of contractors' actual costs – helpful when needing to settle a dispute. But the approach to resolving

---

[14] Institution of Civil Engineers, Crossrail project: commercial aspects of works contracts for London's Elizabeth line, November 2017
[15] Institution of Civil Engineers, Crossrail project: commercial aspects of works contracts for London's Elizabeth line, November 2017

compensation events was not always straightforward and depended on Crossrail, its programme and project partners and contractors adhering to the processes set out in the contract.

For the larger contracts several of the bidders formed joint ventures of two, three or four companies. This reduced the risk for individual contractors as well as giving them a pool of expertise and experience to draw from. But for Crossrail there could be drawbacks to this approach: joint ventures could take time to 'bed down' and establish common systems and processes. Some joint ventures were said to struggle with internal governance and decision-making.[16]

## Tunnelling contracts

The Crossrail tunnelling strategy is explained in chapter 4. Three major contracts were let for tunnelling and included a wide-ranging remit including procurement of tunnel boring machines and establishing concrete segment manufacturing facilities.

Contracts were also let for work on station platform and passageway tunnels with packages combined and revised as the approach to tunnelling developed. C410 sprayed concrete lining works at Tottenham Court Road were delivered by BFK alongside the western running tunnels package. C510 saw a consortium of Alpine BeMo, Balfour Beatty, Morgan Sindall and Vinci deliver early access shafts and sprayed concrete lining works at Liverpool Street and Whitechapel.

---

[16] Institution of Civil Engineers, Crossrail project: commercial aspects of works contracts for London's Elizabeth line, November 2017

# Tunnel contract awards

| Contract number | Name | Awarded to |
|---|---|---|
| 300 | Western running tunnels | BAM Nuttall/Ferrovial/Kier |
| 305 | Eastern running tunnels | Dragados/Sisk |
| 310 | Thames Tunnel | Hochtief/Murphy |
| 315 | Connaught Tunnel | Vinci |
| 340 | Victoria Dock Portal | Vinci |
| 350 | Pudding Mill Lane Portal | Morgan Sindall |
| 360 | Mile End and Eleanor Street shafts | Costain/Skanska |
| 410 | Station tunnels – Bond Street/Tottenham Court Road | BAM Nuttall/Ferrovial/Kier |
| 420 | Station tunnels – Tottenham Court Road | Subsumed into C410 |
| 500 | Station tunnels – Liverpool Street | Subsumed into C510 |
| 510 | Station tunnels – Liverpool Street/Whitechapel | Alpine BeMo/Balfour Beatty/Morgan Sindall/Vinci |

# Crossrail station contracts

Design and enabling works cleared the way for the construction of the Elizabeth line stations. For the central London stations Crossrail let advance works packages which in most cases consisted of piling and diaphragm wall works. These were followed by prestigious major station packages which in some cases would be worth hundreds of millions pounds. Contract values can be found in chapter 11.

Each of the central London stations was a significant project and Crossrail was unusual in the number of major schemes it grouped together. In 2017 the combined value of the eight main station contracts was put at £2.3 billion. The Farringdon contract, C435, differed from the others in that it included the sprayed concrete tunnel enlargements at the station as well as the station construction and fit-out.[17]

## Shortlists for main station contracts

| Contract number | Station | Shortlisted companies – winner shown in bold |
|---|---|---|
| 405 | Paddington | Balfour Beatty/Morgan Sindall/Vinci, **Costain/ Skanska**, Laing O'Rourke/Strabag, BAM Nuttall/ Ferrovial Agroman/Kier, Carillion |
| 412 | Bond Street | Laing O'Rourke, **Costain/Skanska**, Lend Lease, BAM Nuttall/Ferrovial Agroman/Kier |
| 422 | Tottenham Court Road | Lend Lease, Dragados/Sisk, **Laing O'Rourke**, BAM Nuttall/Ferrovial Agroman/Kier |
| 435 | Farringdon | Balfour Beatty/Alpine BeMo/Morgan Sindall/Vinci, Costain/Skanska, Laing O'Rourke/Strabag, **BAM Nuttall/Ferrovial Agroman/Kier** |
| 502 | Liverpool Street | Balfour Beatty/Morgan Sindall/Vinci, Costain/ Skanska, **Laing O'Rourke**, Dragados/Sisk |
| 512 | Whitechapel | **Balfour Beatty/Morgan Sindall/Vinci**, Costain/ Skanska, Dragados/Sisk, BAM Nuttall/Ferrovial Agroman/Kier |

## Station contract awards

| Contract number | Name | Awarded to |
|---|---|---|
| 405 | Paddington main works | Costain/Skanska |
| 411 | Bond Street advance works | Costain/Skanska |
| 412 | Bond Street main works | Costain/Skanska |

(Table continued)

---

[17] Institution of Civil Engineers, Crossrail project: commercial aspects of works contracts for London's Elizabeth line, November 2017

(Table continued)

| Contract number | Name | Awarded to |
|---|---|---|
| 421 | Tottenham Court Road advance works | Balfour Beatty/Morgan Sindall/Vinci |
| 422 | Tottenham Court Road main works | Laing O'Rourke |
| 430 | Farringdon advance works | Laing O'Rourke/Strabag |
| 435 | Farringdon main works | BAM Nuttall/Ferrovial/Kier |
| 501 | Liverpool Street advance works | BAM Nuttall/Kier |
| 502 | Liverpool Street main works | Laing O'Rourke |
| 503 | Liverpool Street advance works | Vinci |
| 511 | Whitechapel advance works | BAM Nuttall/Kier |
| 512 | Whitechapel main works | Balfour Beatty/Morgan Sindall/Vinci |
| 520 | Custom House main works | Laing O'Rourke |
| 530 | Woolwich box and portal fit-out | Balfour Beatty |

## Systems contracts

For tunnelling and station construction the approach chosen by Crossrail Ltd was to let a series of design contracts. Construction packages were awarded separately with the successful bidders handed the output of the design work which gave them a detailed specification to work to.

With railway systems Crossrail took a different approach with contractors given responsibility for design as well as delivery; the aim was for the suppliers to come up with the best ways of meeting Crossrail's requirements. In 2014 Crossrail technical director Chris Sexton explained to me: "We provide the specifications and in some cases a reference design but then it's for the supply chain to provide back to us the completed design which meets the functional specifications which we have set."[18]

---

[18] Modern Railways, Crossrail systems feature, September 2014

Six key systems contracts were ultimately let by Crossrail; more were planned but consultation with suppliers led to a decision to merge some contacts. C610 was the principal Crossrail systems contract, an overarching package won by a joint venture of Alstom, Costain and Travaux Sud Ouest. TSO had previously been involved in the French LGV Est and LGV Atlantique high speed rail projects and worked in partnership with Alstom during construction of the second section of the Channel Tunnel Rail Link, now High Speed 1.

Once the Crossrail tunnels were built they were handed over to the C610 team which had a wide-ranging remit; this included installing 50 ventilation fans, 40km of walkways, 60 drainage pumps, 30km of fire mains as well as lighting throughout the entire length of tunnels. ATC (Alstom/TSO/Costain) would operate a 465 metre long concrete train as part of C610.

TSO was responsible for the initial services including tunnel lighting, fire mains, gravity and pumped drainage, tunnel walkways, ventilation and cross passage doors. The company fitted overhead line equipment and the contract also included the installation of standard and floating track slab. System-wide low voltage power distribution and logistics were included in C610.

Alstom and Costain, this time without TSO, were charged with delivering the power required by the new infrastructure and trains. Under C644 the joint venture provided a 400V supply to feed the overhead wires attached to fixed beams throughout the Elizabeth line tunnels. For Crossrail the rigid overhead conductor system was used for the first time in the UK to power heavy rail main line trains through tunnels. The contract also included electro-magnetic computability testing, earthing and bonding, and providing auto transformer feeder stations.

The other power supply contract, C650, was for the high voltage distribution system that supplies all Crossrail stations, shafts and portals. Alstom and Costain were responsible for supplying and installing 22KV/415V and 11KV/415V transformers, cables and switchgear.

Signalling for the Crossrail central operating section was contracted to Siemens under C620. Siemens bid for the package with Invensys Rail which the German company then acquired in 2013. For the central section of railway Siemens installed the radio-based control system Trainguard MT with automatic train operation (ATO), the operations control system Vicos and radio transmission system Airlink, including the integration between ETCS, TPWS and CBTC. An explanation of the Elizabeth line signalling system can be found in chapter 8.

Under C620 Siemens was responsible for the signalling interfaces with Network Rail. It also designed permanent test facilities for the Old Oak Common depot as well as the test track systems that were used prior to the new Class 345 trains entering service. The contract included signalling control at the rail control centre and the development of the notified national technical rules (NNTR) which are used to ensure safe, efficient and sustainable railway operation.

Siemens was also awarded the C660 communications package which supports tunnel signalling as well as other essential railway systems. This includes the GSM-R voice and data link used by train drivers, the Tetra digital radio used by the emergency services, operational telephones, station management systems, access control and intruder detection. C660 covered CCTV, customer information systems and passenger help points. Siemens was also responsible for the supervisory control and data acquisition (SCADA) diagnostics equipment that is used to monitor station and tunnel systems.

Main power from the electricity network to the Elizabeth line is provided via bulk power supply points at Kensal Green and Pudding Mill Lane and the railway can operate even if one of these supply points fails for any reason. Crossrail contracted London Underground to deliver new station control rooms now used at stations served by both Crossrail and Tube lines.

For lifts and escalators Crossrail worked with TfL to let a contract which could also cover the installation of new systems at London Underground stations – an approach designed to keep costs down.

In total 81 escalators were specified for nine new Elizabeth line stations between Paddington and Woolwich with a total length of

2.8 kilometres. Crossrail/TfL supplier Otis (C740) provided 64 of these with Kone responsible for the 17 at Canary Wharf. There are 54 lifts at these nine stations. The longest escalator on the Elizabeth line is at the Bond Street eastern ticket hall (60 metres) and the shortest (18.5 metres) at the Liverpool Street Broadgate entrance. Escalators in the new Elizabeth line stations travel at a speed of 0.75 metres per second with the exception of Canary Wharf where nine of the 17 escalators run at 0.65 metres per second due to a lower escalator rise and the space available for installation. These nine escalators also have one less flat step at the top.

## System contract awards

| Contract number | Name | Awarded to |
|---|---|---|
| 610 | Track, overhead line electrification and logistics | Alstom, TSO, Costain |
| 620 | Signalling Systems | Invensys (Siemens) |
| 630 | Tunnelling, mechanical and electrical systems | Subsumed into C610 |
| 631 | Platform screen doors | Knorr Bremse |
| 641 | Kensal Green bulk supply point | National Grid |
| 643 | Pudding Mill Lane bulk supplypoint | National Grid |
| 644 | Traction power | Alstom/Costain |
| 650 | Non-traction power (22KV and 11KV) | Alstom/Costain |
| 651 | Limmo bulk supply point | UK Power Networks |
| 660 | Communication and control systems | Siemens |
| 670 | Control/SCADA systems | Subsumed into C660 |
| 680 | Radios/data transmission | Subsumed into C660 |
| 701 | Instrumentation monitoring | Instrumentation Testing and Monitoring |
| 730 | Lifts | Kone |
| 740 | Escalators | Otis |

## Other tier one contracts

| Contract number | Name | Awarded to |
|---|---|---|
| 336 | Paddington New Yard (Westbourne Park elevated bus deck) | Costain |
| 695 | Plumstead maintenance facility | Alstom/TSO/Costain |
| 806 | Wallasea temporary jetty | BAM Nuttall |
| 807 | Marine transportation | BAM Nuttall/Van Oord |
| 815 | Tunnelling academy | VolkerFitzpatrick |
| 816 | Ilford Yard depot intrusive survey | VolkerFitzpatrick |
| 828 | Ilford Yard stabling sidings | VolkerFitzpatrick |

## Prime positions

The tier one contracts listed in this chapter provide a useful overview of *what* was required to build and deliver the Elizabeth line and we have also seen, from the contractors appointed, *who* would be tasked with making this happen.

On one hand this is a complicated story of mobilisation – going from nothing to having hundreds of bids scrutinised, contracts awarded and an army of suppliers and specialists on standby. The contract packages are fascinating when you start scratching beneath the surface and realising that behind each award there's a hidden world of smaller tier two, three and beyond sub-contractors who found new opportunities and revenue streams from the Crossrail programme.

On the other hand, when you focus on the high-value contracts listed in bold, responsibility for this multi-billion pound railway venture can be seen to rest in the hands of a small number of trusted partners. Between twenty and twenty-five companies would be pivotal to Crossrail. Large sums of the budget would be channelled to these firms who in return would create structures and systems that most people barely understand. How they would do that is explored in the next chapter.

Canalside view of Carillion activity. The Paddington
Integrated Project takes shape with the new Hammersmith &
City line ticket hall, a new entrance to Paddington station and,
behind the hoardings at the back of the picture, the new station
taxi rank, relocated from Departures Road to allow
construction of the new Elizabeth line station
by a joint venture of Costain and Skanska.

# 4. THE BUILD

Crossrail was, fundamentally, a scheme to build new railway tunnels under London. There are many other complex, fascinating facets of the project but without the tunnels you have no Elizabeth line.

So once the plans for Crossrail were signed off and the construction programme mobilised the immediate priority was how to deliver 42 kilometres of new railway tunnels. This would transform the scheme into a major civil engineering venture, for a time Europe's largest rail infrastructure project.

To understand how it was done one needs to be familiar with the geography of the route. The basic 'cross-rail' concept links existing London rail termini west (Paddington) and east (Liverpool Street) and by pathing commuter services into new tunnels releases capacity by doing away with the need for terminating platforms (Elizabeth line trains don't terminate in the centre[1]) as well as putting central London destinations that people want to reach on a 'proper' (sorry, Underground) railway line.

Being a little more specific then, the tunnels start in west London, just under a kilometre west of Paddington, at the Royal Oak Portal.

From here we have two train tunnels, one for each direction of travel, running all the way under central and east London to a rather imposing set of caverns at Stepney Green where the Elizabeth line route divides. Taking the north eastern branch we continue in tunnel for a generous two and a half kilometres before reaching the surface at the Pudding Mill Lane Portal, just outside the Queen Elizabeth

---

[1] Although when the Elizabeth line opened a small number of services started and terminated from Liverpool Street high-level station platforms 16 and 17.

Olympic Park stadium, main venue for the London 2012 Olympic Games.

The branch to the south east is more complex. Trains head under the Docklands business district, travelling through Canary Wharf station before emerging at Victoria Dock Portal. There's a short section at surface level before we head under the docks via the Connaught Tunnel. This is not a new Crossrail tunnel, it's an historic artefact built in 1878 which, prior to Crossrail, served as a workhorse of Docklands industry before enjoying a more sedate role as part of the North London line route to North Woolwich, closed in 2006. While certainly not new-build the Connaught Tunnel had to be painstakingly modernised, with the tunnel profile enlarged, to form part of the Elizabeth line.

## Tunnel drive details

| Drive | TBM | Launch site | Destination | Finish date | Total tunnelling days including delays | Average rings per day |
|---|---|---|---|---|---|---|
| X | Phyllis | Royal Oak | Farringdon | 7/10/13 | 519 | 9 |
| X | Ada | Royal Oak | Farringdon | 27/1/14 | 522 | 8.9 |
| H | Sophia | Plumstead | North Woolwich | 29/1/14 | 392 | 4.4 |
| Z | Jessica | Pudding Mill Lane | Stepney Green | 13/2/14 | 172 | 9.8 |
| H | Mary | Plumstead | North Woolwich | 22/5/14 | 367 | 4.6 |
| Z | Ellie | Pudding Mill Lane | Stepney Green | 13/6/14 | 127 | 13.9 |
| G | Jessica | Limmo | Victoria Dock | 5/8/14 | 42 | 12.4 |
| G | Ellie | Limmo | Victoria Dock | 18/10/14 | 69 | 7.6 |
| Y | Elizabeth | Limmo | Farringdon | 10/5/15 | 896 | 5.1 |
| Y | Victoria | Limmo | Farringdon | 26/5/15 | 895 | 5.5 |

Out of the Connaught, and back above ground, Crossrail trains continue to the North Woolwich Portal where they go underground once more, through the new Thames Tunnel, which takes us, unsurprisingly, under the River Thames, through Woolwich, and then emerges at the Plumstead Portal, allowing trains to continue on to Abbey Wood.

With the geography lesson complete the basic Crossrail tunnelling plan came down to this: Five tunnel portals would be required with three of them connected by twin east and west running tunnels that would meet at Farringdon – the hub of the Elizabeth line tunnelling enterprise. The other two tunnel portals would be joined by a shorter, twin bore, Thames Tunnel. In addition, the existing Connaught Tunnel would need to be extensively refurbished. To construct the 42km of new-build tunnels (2 x 21km – about 26 miles in total) eight tunnel boring machines would be purchased to deliver a carefully choreographed sequence of tunnel drives.

## Tunnel drives

There were two main tunnelling drives, west and east, labelled by Crossrail as X and Y. The 6.16km (3.8 miles) western drive went from the Royal Oak Portal to Farringdon. TBM Phyllis completed the first tunnel bore with sister machine Ada responsible for the second.

Although the western drive, X, began first, the longest tunnelling machine journey would be from the east. TBMs Victoria and Elizabeth began tunnelling from the Limmo Peninsula work site (beside a curve of the Thames near Canning Town station) in November and December 2012 respectively. Like the western machines, they were destined for Farringdon once they had completed 8.3km (5.2 miles) of twin bore tunnels.

The other tunnelling drives were shorter. The north eastern branch of the new underground railway was constructed through drive, Z, which would create the 2.7km (1.7 miles) of twin bore tunnels linking the Stepney Green junction (through which Victoria and Elizabeth had already passed) with the Pudding Mill Lane Portal.

Launched from Pudding Mill Lane on 16 August 2013, TBM Jessica would complete the first bore with sister machine Ellie responsible for the parallel route. Upon arrival at Stepney Jessica and Ellie were dismantled and wheeled through the Drive Y tunnels to Limmo and moved into position at the bottom of the shaft ready to construct the final two Crossrail tunnels between Limmo and Victoria Dock portal. This Drive G would create the shortest of the Crossrail tunnels with each bore 900 metres long.

In south east London TBMs Mary and Sophia were responsible for Drive H between the Plumstead and North Woolwich portals. Crossrail's two slurry TBMs would head through the Woolwich station box before constructing the new Thames Tunnel. Unlike the other tunnel drives where the TBMs dug out London clay, Sophia

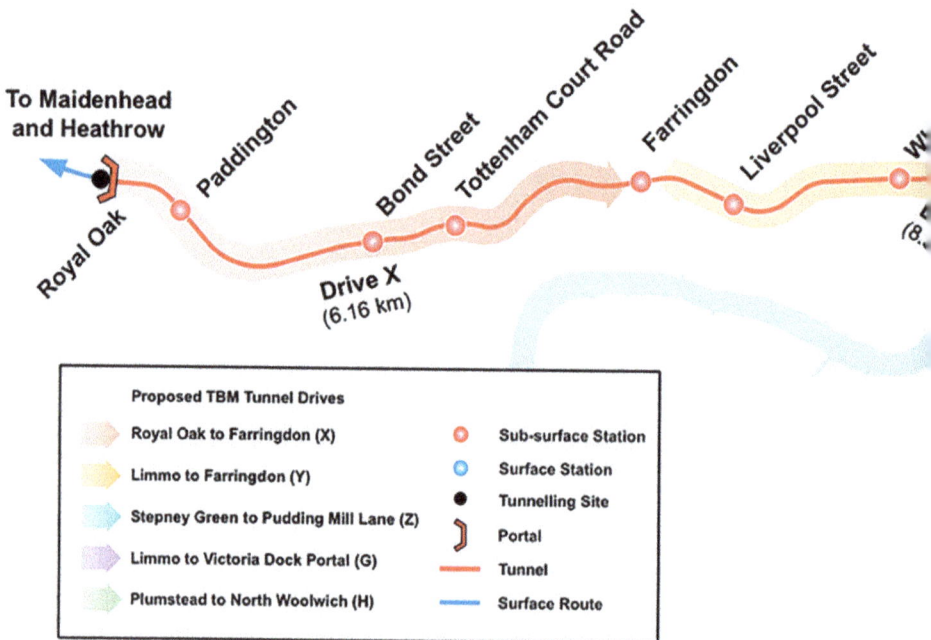

**Proposed TBM Tunnel Drives**

Royal Oak to Farringdon (X)

Limmo to Farringdon (Y)

Stepney Green to Pudding Mill Lane (Z)

Limmo to Victoria Dock Portal (G)

Plumstead to North Woolwich (H)

○ Sub-surface Station

○ Surface Station

● Tunnelling Site

⌡ Portal

— Tunnel

— Surface Route

and Mary were equipped to bore through chalk, flint and gravel with this liquid slurry being piped out of the Plumstead portal to a treatment plant for conversion to chalk 'cakes'.

## Tunnel drives, distances and contractors

| Drive | Tunnel | Distance (km) | Contractors |
|-------|--------|---------------|-------------|
| G | Eastern | 0.93 | C305 Dragados/Sisk |
| H | Thames | 2.64 | C310 Hochtief/Murphy |
| X | Western | 6.16 | C300 BAM/Ferrovial/Kier |
| Y | Eastern (main) | 8.30 | C305 Dragados/Sisk |
| Z | Eastern | 2.72 | C305 Dragados/Sisk |

## Naming the TBMs

Following tunnelling tradition the Crossrail tunnel boring machines were named after women. Names were chosen in a series of competitions – Crossrail nominated the names of people with local or historical significance and then schoolchildren had the final vote.

The names of the TBMs were decided in pairs. Pioneering women Phyllis Pearsall, creator of one of the first London A-Z maps, and Ada Lovelace, the early computer scientist who realised that machines could be programmed to do different things, gave their first names to TBM1 and 2 (Drive X). TBMs 3 and 4 (Drive Y) were named after English queens Victoria and Elizabeth.

For TBMs 5 and 6 (Drive H) there was recognition of the wives of the famous Brunel railway engineers – Mary, wife of Isambard Kingdom Brunel, whose credits included the Great Western Railway, and Sophia, mother of Isambard Kingdom and wife of Marc Brunel, who started the first tunnel to cross under the River Thames in 1825. With the Crossrail build overlapping with the 2012 Olympic Games, Olympic heptathlete Jessica Ennis-Hill and Paralympic swimmer Ellie Simmonds saw their names used for TBMs 7 and 8 (Drives Z and G), particularly appropriate as these machines were initially launched from the Pudding Mill Lane portal, close to the Queen Elizabeth Park, the main venue for the Games in east London.

Other names from different spheres of life and fiction were considered but ultimately rejected as names for Crossrail TBMs. These included Betsy and Dora – Characters from Charles Dickens' novel David Copperfield; Nancy and Nell – from Dickens' Oliver Twist and The Old Curiosity Shop; Gracie [Fields] and Vera [Lynn] – for epitomising the Blitz spirit; and Dorothy and Audrey – after 1948 London Olympic medal-winning British sprinters Dorothy Manley and Audrey Williamson. Also rejected were the BBC Eastenders-inspired names Pat and Peggy.

## Royal Oak Portal

Before reporting on Crossrail progress I had never given much thought to how tunnel boring machines actually break ground (in cartoons they just point downwards and get going). I would soon learn that portals where not just found in science fiction and were a fundamental part of the tunnelling process.

Work on the Elizabeth line's western tunnel portal, at Royal Oak in west London, began in January 2010. Initially it consisted of a 285 metre long ramp down to a concrete wall built in a 21 metre wide corridor sandwiched between the A40 Westway and the Underground and main line railway out of Paddington.

To enable Elizabeth line tracks to progress from ground level into the tunnels the portal required the construction of a retained cutting using sheet piles driven into the ground on each side with the ground between excavated to create a ramp. As the slope headed deeper underground diaphragm walls were formed: this is a technique that was widely used across Crossrail sites and involved the construction of concrete walls using a rig that cut a series of overlapping slots in the ground. These slots were then filled with bentonite – like thick sandy water – to stop the sides collapsing. Steel reinforcement rebar cages were lowered into position through the bentonite and then the slot filled with concrete from the bottom upwards. As the displaced bentonite rose to the top it left a reinforced concrete wall embedded in the ground.

Fifty-four steel and concrete props supported the portal walls during the excavation and when tunnel boring began a concrete deck was installed just outside the entrance to enclose the steel props visible during early construction.

At the deepest end of the cutting a concrete headwall was constructed and a lining wall with a pair of 7.24 metre diameter tunnel eyes installed on to it – targets for the TBM cutter heads. These would incorporate steel rings and support the tunnel opening when the TBMs broke through the headwall, helping protect TBM cutter heads from damage.

The site was handed over to BAM Nuttall/Ferrovial Agroman/Kier (C300) in autumn 2011, allowing the main tunnelling contractor to mobilise and construct a temporary railway that would be used to transport the TBMs from the assembly site at Westbourne Park and concrete tunnel segments from Old Oak Common to the tunnel eyes. BFK would also sink three rows of piles on the Paddington side of the portal which would act as supporting walls until the tunnels were deep enough underground not to require any additional strengthening.

Amid all the modern engineering paraphernalia one contrasting sight at Royal Oak, fixed between the two tunnel eyes of the portal, was a case containing an icon of St Barbara, the patron saint of miners. I'm sure I'm not alone in assuming that, because they happen, all modern engineering activities must be safe and forgetting that, even with modern technology, tunnelling underground carries an inherent risk of danger. I would spot the St Barbara icon at other Crossrail tunnelling sites as the construction programme continued, an unusual insight into the thoughts and fears of outwardly indomitable workers as they headed to work underground.

## OOC segment factory

Crossrail catalysed the transformation of the Old Oak Common site in west London which went through a series of changes, some temporary, some permanent, during the project. It is now the location for the main Elizabeth line train depot, a two-storey building, with room to accommodate up to nine trains at once, which may in future be redeveloped as part of High Speed 2. But early in the tunnelling process it would have a key role as a marshalling yard for the 74,000 concrete segments required by the western tunnelling machines heading for Farringdon.

Old Oak Common has a long history of train building and maintenance. To prepare for tunnelling C300 contractor BFK (BAM Nuttall/Ferrovial Agroman/Kier) spent £4.5 million converting the former DB Schenker/EWS depot site at Old Oak, last used in 2007 and approximately three miles west of the Royal Oak Portal, into a concrete segment manufacturing facility.

Moulds at Old Oak Common were used to cast the
concrete segments required to line Crossrail tunnels.

Around £3 million was spent buying 213 concrete moulds from
France's CBE Group and these were housed in a newly built hangar,
costing a further £1 million. Tunnelling is a dirty job and as the
TBMs fed back excavated material from the cutter heads the portal
approaches would be dominated by heaps of brown clay collected
from conveyor belts. In contrast, the segment manufacturing facility
was one of the most colourful industrial sites I have witnessed with
stacks of moulds red, blue, yellow and green according to the side
and size of each piece required to form one of the 6.2 metre diameter
tunnel rings. Five smaller white moulds were used to make the
segments for platform connecting tunnels – a key part of station
tunnels package C410.

A further £500,000 was spent by BFK converting the old train
yard – the last of three turntables on the site had been removed in
2011 – into a vast area of gravel and concrete stripes. Segments were
stockpiled on the gravel and the concrete roads in between allowed

lorries to quickly collect segments for transport to Westbourne Park. Plans to move the segments by rail from Old Oak to Westbourne Park were rejected because there was insufficient space both to bring in segment trains and have trains taking spoil out. A secondary idea of using canal barges was also dropped because it would have turned the Grand Union Canal into a construction site for two years and required major dredging work. Building two sidings at Westbourne Park for trains to come in and collect tunnel spoil proved the most pressing need and it was decided the segments would be moved the short distance to the tunnel approach by road.

Old Oak Common functioned as a concrete segment manufacturing facility for around two years. After this BFK was contracted to demolish the hangar which housed the moulds and remove all concrete paving to allow construction of Crossrail's main train depot. Planning permission for the depot was granted in December 2011.

Between Old Oak and the Royal Oak Portal everything needed for tunnelling was marshalled at Westbourne Park. The first two of Crossrail's eight tunnel boring machines were assembled here ready to be wheeled to the portal; squeezed between the Great Western/ London Underground sub-surface lines and the A40 there was not enough space to assemble the machines closer to the portal.

Before tunnelling started engineers assembled a series of Meccano-like metal frames with rollers that supported twin conveyor belts to move tunnel spoil from the back of each TBM to freight wagon trains waiting at Westbourne Park. As the TBMs for each bore progressed the conveyors were extended – outside the portal they hung from specially erected gantries and in tunnels on brackets fixed to the concrete segments forming the new tunnel walls.

## Starting out (May 2012)

On 3 May 2012 Crossrail tunnelling began with the launch of the first tunnel boring machine close to the Royal Oak Portal in west

London. Ahead of this, at a media event on 13 March, Transport Secretary Justine Greening and Mayor of London Boris Johnson pressed a button to start Crossrail's first tunnel boring machine, Phyllis. Following the event, Phyllis was wheeled 400 metres from the assembly site at Westbourne Park to Royal Oak Portal. Hampden Street footbridge had to be jacked up to enable the 7.1 metre diameter machine to pass underneath and before tunnelling could start a launch structure needed to be built at the portal (as described above). This allowed the TBM to use its hydraulic rams to thrust itself forward into the ground. A steel seal was fitted around the portal entrance to support the ground during the early stages of tunnelling.

By this point procurement for the tunnelling contracts was complete. The biggest – running tunnels east C305 – went to a joint venture of John Sisk and Dragados which took charge of creating tunnels as far west as Farringdon. The western section, between Farringdon and Royal Oak, was covered by C300 which was awarded to Team BFK – an alliance of BAM Nuttall/Ferrovial Agroman/Kier and it was this outfit that got tunnelling first with the launch at Royal Oak.

The four drives these contractors were responsible for required the use of earth pressure balance tunnelling machines. The third tunnelling contract, Thames Tunnel package C310, was awarded to a joint venture of Hochtief and Murphy. As this involved boring under the river through chalk rather than the London clay, sand and gravels encountered along most of the other Crossrail tunnel routes, slurry tunnel boring machines would be required here.

The contractors would need some heavy machinery to actually dig the tunnels and it fell to German company Herrenknecht's Schwanau factory, just across the border from Strasbourg, to supply the kit. After making much of the project's buy British credentials Crossrail justified shopping abroad by the absence of any TBM supply options in the UK. This perhaps reflects the scant number of tunnels built in the UK prior to Crossrail and High Speed 2; smaller tunnelling enterprises were required for the London Underground Jubilee line extension and Docklands Light Railway extension to Lewisham, both completed in 1999. Each Crossrail TBM would cost approximately £10 million; the components were shipped from Germany to Tilbury Docks ready for assembly in the UK.

At the Royal Oak TBM launch I was able to see the first of the Herrenknecht TBMs up close; these 980 tonne beasts were massive but perhaps not as you might imagine them. At the front there was the circular cutting disk, seen on standard TBM press pictures, but behind this was an elaborate – and lengthy – train of machinery, often referred to as the back-up. This meant each TBM was 148 metres long and included pipes and conveyors moving spoil to the rear of the machine and grout-like fillers to the front, space for the 'tunnel gang' – typically 20 people (12 on the TBM itself and eight working from the rear of the machine to ground level) – and a loading mechanism for concrete tunnel segments, which needed to be moved from the back of the machine to the front in just the right order. Each TBM had its own toilets and kitchen.

Functional yes, but outside the tunnel the appearance of this back-up lacked the obvious sophistication we expect from German engineering. The succession of tunnelling trolleys looked more Heath Robinson than Mercedes. As the (earth pressure balance) TBM advanced the cutter head rotated to excavate material which was removed from the cutter head by a giant screw thread and deposited on to a series of conveyor belts, feeding the clay through the TBM, and ultimately out of the tunnel portal. Above ground this unwieldy mass of machinery looked like some industrial relic but as Phyllis began her journey underground it would soon be hidden from view.

As the TBMs advanced forward – Phyllis would ramp up to an average tunnelling rate of around 100 metres a week[2] – precast concrete segments were built in rings behind the machines. The 300mm thick segments, reinforced with steel and polypropylene fibres and designed to last for at least 120 years, could be positioned with millimetre accuracy using a lining segment erector. Once in

---

[2] Keith Sibley, Crossrail area director west

place, the TBM pushed itself off from the last installed segment ring using hydraulic thrust cylinders arranged in a ring.

Crossrail set up two concrete segment manufacturing facilities – Old Oak Common for the western tunnels and Chatham for the eastern tunnels. The segments for the Thames Tunnel (C310) were supplied by Shay Murtagh in County Westmeath, Ireland.

A single tunnel ring was made up of eight concrete pieces; in total 250,000 tunnel segments were required to line the 42 kilometres of Crossrail tunnels. To get the segments to the tunnelling faces narrow gauge railways were installed in the tunnels with small locomotives used to move the tunnel segments to the advancing tunnel boring machines. The locos were also used to transport personnel to and from TBM shifts – an increasingly long commute from tunnel portals as the TBMs advanced.

For users of the Great Western route and London Underground between Royal Oak and Westbourne Park the activity at the tunnelling face could be measured by the quantity of clay that could be seen daily moving along the conveyor belt from the portal to the train sidings. On 27 September 2012 part of the conveyor system collapsed with a hopper falling into one of the waiting train wagons. It made for a dramatic photograph but there were no reported injuries and the conveyor was soon repaired. Later in the programme a Crossrail-led review noted that the project's conveyor systems suffered from several failures, including structural issues.[3] It said conveyor designers for future projects should be made aware of the clogging potential of London clay.

Following the launch of Phyllis at the Royal Oak Portal TBM number two, Ada, was assembled at Westbourne Park and moved to Royal Oak ready to start work on the second bore of the western tunnels once Phyllis had progressed underground. When Ada

---

[3] Institution of Civil Engineers, Crossrail project: machine-driven tunnels on the Elizabeth line, London, May 2017, Crossrail Learning Legacy, Machine driven tunnels on the Elizabeth line

reached Paddington both machines were moved forward through the Paddington station box and then onwards until they had completed the 6.4km drive to Farringdon.

## Eastern front (November 2012)

A critical part of Crossrail's tunnelling programme was its worksite at Limmo Peninsula, situated beside a kink in the Thames.

The triangular site, bounded by the tidal waters of Bow Creek to the west, Canning Town station to the east and the Lower Lea Crossing road bridge to the south, was selected because it allowed the river to be used for delivering materials and also for shipping away material excavated from the new tunnels.

To launch TBMs at Royal Oak involved wheeling the 148 metre long contraptions down a ramp to the newly built portal but at Limmo the machines would start drilling from deep underground. To make this possible two 40 metre deep shafts were dug down which the TBMs were lowered into their start positions for Crossrail's longest tunnelling drive heading east towards Farringdon.

A main, 30 metre diameter shaft would become a permanent Crossrail structure providing ventilation and an emergency escape route once Elizabeth line trains started running. Before this TBM cutter heads for Victoria and Elizabeth were lowered into the shaft to be attached to back up sections of the tunnelling machines.

Rudimentary maths tells us that space at the bottom of the shaft was not large enough to assemble the full 148 metre back up and this is where the Limmo auxiliary shaft assisted: by mining twin tunnels between the shafts Crossrail's tunnelling team created the space needed to put TBMs Victoria and Elizabeth into their starting positions for the 8.3km Drive Y.[4]

---

[4] In 2014 Limmo would also serve as the starting point for Drive G heading in the opposite direction towards Victoria Dock Portal

At Limmo eastern tunnels contractor Dragados/Sisk (C305) had around 500 people working at the shaft as tunnelling got underway. The two shafts were so deep that workers had to carry emergency breathing apparatus with them while underground in case of a gas leak. While you could take the (very high) scaffolding stairs from top to bottom of the shaft the quickest way up or down was to stand in what was effectively a large metal bucket which could be hoisted by a crane.

On the surface the TBM back up components were positioned ready to be lifted down and concrete platforms had been cast to support the 120,000 tunnel segments that would be shipped from the new manufacturing facility that was by now up and running at Chatham. Down underground concrete structures for the TBM launch sites were prepared with tunnel eyes similar to those seen at Royal Oak.

Tunnelling from Limmo began on 29 November 2012 with the launch of TBM Elizabeth – who would travel 1.5km before breaking through into the completed Canary Wharf station box in 2013. Victoria followed to create the second bore. After being wheeled though the station box the machines continued their journey on to Farringdon via Stepney Green, the intersection between Crossrail routes from Shenfield and Abbey Wood.

Once again considerable thought went into how to dispose of the vast quantity of material dug out by the TBMs. A new jetty was built at Instone Wharf near Canning Town, east London, which enabled material excavated from the eastern tunnels to be loaded on to barges, avoiding the need for 30,000 lorry journeys. Around 1.2 million tonnes of tunnel spoil was transported by boat to Wallasea Island in Essex to create a new 1,500 acre RSPB nature reserve. Here BAM Nuttall built a receiving jetty made up of two 800 tonne steel pontoons, each 15 metres wide and able to accommodate two 90 metre, 2,500 tonne ships simultaneously.

At peak times two unloading machines per pontoon serviced four ships and unloaded up to 10,000 tonnes of tunnel spoil over a 24 hour period. The jetty received around 4.5 million tonnes of material excavated during Crossrail's tunnelling operations across London.

For those times when the tide was too low or no ship was waiting at the Instone Wharf jetty, material excavated from tunnels was stored in a 100 x 9 metre muck pit which was dug at Limmo. Spoil from the pit, or direct from the Limmo shafts, could then be moved on a high angle conveyor to the jetty.

## Farringdon arrival (October 2013)

Tunnel boring machine Phyllis arrived at Farringdon on 7 October 2013. Ada followed on 27 January 2014, completing Drive X from the Royal Oak Portal, a journey that had taken the two machines about a year and a half. For those of us familiar with a London Underground map that had changed little in decades this was quite something. And given the lengthy saga of securing permission to start building, the creation of an all-new underground rail tunnel via Paddington, Bond Street and Tottenham Court Road in a matter of months seemed, to anyone with the slightest interest in railways, a major achievement.

Exactly what it was like for the crews of Phyllis and Ada en-route to Farringdon we can only imagine. Even with Herrenknecht's latest tunnelling kit the daily toil in cramped conditions away from daylight and with limited ventilation required considerable commitment. And by the final stages of Drive X it took workers 45 minutes to get from Westbourne Park to the tunnelling face by the narrow gauge railway installed in the recently built tunnels.

During the drives to Farringdon between 14 and 24 tunnel rings a day were installed during three eight hour shifts. Despite the years of careful planning the start of tunnelling came with a certain level of trepidation. Would everything work as planned? Would there be any surprises? In February 2003, during construction of the Channel

Tunnel Rail Link, gardens for three homes in Stratford above the tunnel route collapsed.

Ground movement was a major concern for Crossrail and so, to monitor any changes, thousands of studs and theodolite prisms were fixed to any structure vaguely near Crossrail activity. These fed back a vast quantity of data and were configured to trigger alerts if unexpected ground movement occurred. To further control ground movement, the rate and volume of excavated material from each TBM had to be kept equal with each machine's advance underground. Pressure sensors carefully controlled this equilibrium while belt scales, a laser scanner and a density meter helped measure the volume of material excavated. GPS positioning data revealing each machine's precise location was fed back to a control room above ground where staff monitored the progress of the TBMs.

The year 2012 was significant not only for being the start of Crossrail tunnelling. For the first time since 1948 the Olympic Games would be held in London and, understandably, organisations across the capital and beyond were being encouraged to develop plans that would ensure the smooth running of the sporting tournament. With the critical Olympic period looming,[5] Network Rail issued a six week moratorium preventing Ada going under the Great Western lines, fearing that any ground movement risked proving disastrous while London was in the international spotlight. However, such was the smooth progress of Phyllis and the depth of data with which Network Rail was furnished that the rail body decided to lift the moratorium and Ada was allowed to proceed during the Games.

While the spectre of a Crossrail catastrophe mid-Games proved unfounded there were plenty more narrow escapades in prospect to keep Crossrail teams glued to the geotechnical data feeds as Phyllis and Ada advanced. Perhaps the most high profile was at Tottenham

---

[5] The Olympic Games ran from 27 July to 12 August with the Paralympics following from 29 August to 9 September

Court Road in September 2012. Here Phyllis was programmed to pass above the two Northern line tunnels at the existing Underground station with a clearance of 853mm. This was necessary to avoid the newly constructed Northern line and Goslett Yard (Crossrail ticket hall) escalator boxes. The 980 tonne machine was said to have cleared these by 1.4 metres – or 300-400mm if you account for the piling. For this delicate operation engineers had predicted that the Northern line would lift by 7mm but in the event this tunnel 'heave' was only 1mm. Steve Parker, Crossrail's construction manager for the western tunnels at the time, said he and a colleague were on one of the Northern line platforms when Phyllis passed through: "We could stand and hear our machine working," he said.

Ada's route was further north but, again, there were some tight squeezes. British Museum station on the Central line was closed in 1933 but a passageway which still provides access to a machine room remains and was only 707mm from the Crossrail alignment. Concerns that foam and polymer used by the TBM could leak into the passageway and potentially run down stairs to the Central line prompted plans to install sandbags and remote cameras. In the event the confidence in ground movement controls resulted in London Underground being satisfied that a before and after inspection would suffice when the TBM passed by at the end of September.

## Slurry machine activity

Between the arrival of TBMs Phyllis and Ada at Farringdon from the west and the arrival of Elizabeth and Victoria from the east, four other machines were used for a series of shorter tunnelling drives.

In early 2013 tunnel boring machine number five, Sophia, and six, Mary, began drive H, a 2.6km journey of just over a year that would take them from Plumstead Portal to the North Woolwich Portal. Unlike the other machines Sophia and Mary were designated slurry TBMs, equipped to bore through chalk, flint and gravel. Specially prepared bentonite mud – the slurry – was pumped to the face of each machine to support the ground while cooling and lubricating the cutting tools. It was also used to remove the excavated material – pumped out of the tunnels through pipes (rather than on a conveyor system) to a slurry treatment plant at Plumstead which cleaned the bentonite before returning it to the TBMs. Excavated material was filtered by the treatment plant into cakes of chalk.

Drive H was the responsibility of a joint venture between Hochtief and Murphy (C310). Following their launch from Plumstead the two machines arrived at the Woolwich station box, excavating nearly 100,000 tonnes of material and installing 811 concrete tunnel rings over a three month period. Sophia and Mary would then continue under the river to complete the Elizabeth line's Thames Tunnel on 22 May 2014.

## Pudding Mill Lane and the Olympics

By autumn 2013 a seventh TBM, Jessica, had started tunnelling from the Pudding Mill Lane Portal, heading for the Stepney Green caverns where the Elizabeth line north eastern and south eastern branches merge. Sister machine Ellie – Crossrail's eighth TBM – started digging the other Drive Z tunnel with both machines arriving at Stepney by the middle of 2014.

Overlooked by the main stadium built for the 2012 Olympic Games, Pudding Mill Lane is where Crossrail's north eastern branch emerges above ground and joins the Great Eastern rail corridor towards Stratford and Shenfield. But to make this connection, within a site hemmed in by the River Lea and Blackwall Tunnel approach to the south west, the City Mill River and Greenway embankment (in which is concealed the Northern Outfall Sewer) to the north east, and the railway to the north west, required considerable rejigging of existing assets.

These included the cables in the Lea towpath supplying Canary Wharf, other cables serving the Queen Elizabeth Olympic Park and a brick barrelled Bazalgette-built sewer described by one site manager as a 'golden asset'. But the most significant piece of work has been the diversion of the Docklands Light Railway, away from the Great Eastern corridor, to create a rail-locked island from where the Elizabeth line could emerge.

For the diversion a new viaduct for the DLR was built by contractor Morgan Sindall (C350), resulting in a DLR-Elizabeth line crossover. New, twin track Crossrail lines, heading out of central London, and bound for Shenfield, now emerge from tunnel portals next to the River Lea with a sub-surface box taking lines gradually upwards, but under the new DLR alignment, and into the island site where a new rail junction with the Great Eastern line has been created.

As Pudding Mill Lane Portal was built it gradually became clear how this would work. Sub-surface tunnel eyes right next to the River Lea marked the point where Crossrail tunnel bores from Stepney Green junction would end. The structure that was subsequently created now also serves as an emergency intervention point with stairs up the 13 metres to a head house at ground level.

The Crossrail tunnels here have some of the steepest gradients on Britain's rail network; this helped to reduce costs, minimise environmental impacts and slot the new railway in between physical constraints. Because of the steep ramps portal approaches include coverings to ensure rail adhesion. Indeed Pudding Mill Lane's actual 'portal' consists of a 300 metre long cut-and-cover tunnel structure and a 120 metre long covered approach ramp which crosses beneath the new DLR viaduct and then finally reaches surface level, joining the Great Eastern corridor on the site of the old Pudding Mill Lane DLR station. This tram stop-like relic, built when the DLR still enjoyed a relatively sedentary existence and so small it had to be closed during the 2012 Games (despite its plum location next to the London Stadium), was demolished to make space.

A four span bridge carries the Elizabeth line over what was the actual Pudding Mill Lane but, due to the restricted headroom for vehicles below, has since been diverted along Marshgate Lane to become the through route under the Great Eastern/Crossrail/DLR corridor. Finally, a track support slab brings Crossrail up to Great Eastern level. While the tunnel openings were visible during construction, now that the cut and cover Pudding Mill Lane structure is finished the actual portal is hidden from view and trains do not emerge into daylight until they cross over Marshgate Lane.

Although rebuilding this section of the DLR was prompted by Crossrail, the project supported Transport for London's plans to increase capacity on the light rail network including double-tracking the single line section between Bow Church and Stratford (via Pudding Mill Lane).

As part of construction of the DLR viaduct a new Pudding Mill Lane station was built, larger and better specified than the one it replaced – indeed the largest built on the DLR network to date. Designed by architect Weston Williamson, the new station opened on 28 April 2014, at the same time as the new DLR viaduct, and includes passive provision for escalators which could be installed if demand from the new homes and businesses planned for the area materialises.

The DLR rejoins the Great Eastern corridor above the Greenway, which as well as housing a sewer serves as a footpath – it formed the main walking route between West Ham station and the Olympic Park during the 2012 Games.

Following construction of the new DLR viaduct a four day engineering possession at Easter 2014 allowed the DLR to switch from old to new track. That in turn allowed hoardings to be put up around the island site, the old Pudding Mill Lane station to be demolished, and piling for Crossrail's final approach to the Great Eastern to get underway.

## More DLR changes

Docklands Light Railway realignment works were also required on the Beckton DLR branch to enable construction of the Victoria Dock Portal. To ensure Crossrail tracks in and out of the eastern running tunnels could emerge at surface level outside Custom House station, but with an acceptable gradient for trains, it was decided the portal needed to be built just to the east of Royal Victoria DLR station.

Plans were worked up for a portal structure and ramp located between Victoria Dock Road and the DLR Beckton branch tracks (linking Royal Victoria and Custom House DLR stations). In order to create sufficient space to construct the portal box it was necessary to realign approximately 420 metres of railway with the alignment moved south. This involved the demolition and replacement of retaining walls as well as alterations to pylons, signalling and control systems.

Just as DLR changes at Pudding Mill Lane were included in Morgan Sindall's C350 PML contract, so the Victoria Dock DLR changes formed part of the C340 Victoria Dock Portal package, contracted to Vinci. The other portals were built under C330 (Royal Oak) and C310 (Thames Tunnel – including Plumstead and North Woolwich portals).

TBMs Jessica and Ellie were commandeered for a second, shorter job to complete the tunnelling programme. From Stepney both machines were wheeled through the south eastern tunnels to the tunnel shafts at Limmo where the eastbound drive Y had begun. Once in position at Limmo the two machines tunnelled the short, 0.93km distance from the shaft to the Victoria Dock Portal with Ellie completing this Drive G in October 2014.

Figures recording the accomplishments of the different TBM crews show that productivity on this Drive G, as well as Drive Z from Pudding Mill Lane, was generally higher than for the longer drives. This is attributed, at least in part, to the experience of the Drive G and Z teams who had already worked together on previous Crossrail tunnelling drives.[6]

---

[6] Institution of Civil Engineers, Crossrail project: machine-driven tunnels on the Elizabeth line, London, May 2017, Crossrail Learning Legacy, Machine driven tunnels on the Elizabeth line

## Second Farringdon arrival and end of tunnelling (May 2015)

At 0530, on 23 May 2015, tunnel boring machine Victoria broke through into Farringdon station and three days later completed the 8.3km Drive Y journey from Limmo Pensinsula. Just over three years after tunnelling began, Crossrail's east and west tunnels had been joined up.

The historic breakthrough meant it would soon be possible to walk under central London, all the way from the Royal Oak Portal to either Pudding Mill Lane or Plumstead – where track to take Crossrail trains on to Abbey Wood station had already been laid. The tunnels might not yet be ready for trains but a drone fly-through provided a glimpse of the route trains and their passengers would follow.

The arrival of TBM Victoria, and sister machine Elizabeth a couple of weeks before, meant main Crossrail tunnelling was finished. Spray concrete lining work to create station tunnels and passageways would continue but installation of the 200,000 plus concrete segments to form the twin running tunnels was done.

Aside from the not insignificant civil engineering milestone of creating 42km (26 miles) of tunnels underneath London, the breakthrough moment symbolised other, perhaps even more significant achievements. As one reporter standing inside – and dwarfed by the scale of – a completed Crossrail tunnel noted, it is incredible that someone chose to make such a colossal tunnel. It is even more incredible that others went ahead and actually built it.

Connecting the western tunnels (constructed by the BAM Nuttall/ Ferrovial/Kier consortium) and the eastern tunnels (built by the Dragados/Sisk joint venture) meant work could now concentrate on fit-out. Concrete slabs to hold track would be cast and eventually new rail and overhead line electrification equipment installed.

Shortly after the breakthrough Prime Minister David Cameron and Mayor of London Boris Johnson headed underground to see for themselves the new railway tunnels at Farringdon with Johnson describing the completion of tunnelling as a "landmark moment".

## TBM decommissioning

Herrenknecht's tunnel boring machines, with their extensive trailer systems, included many parts that could be returned to the manufacturer for reuse on other projects – and allowing the refund of a TBM 'deposit' – if those parts could be retrieved from underground.

Following their arrival at Farringdon TBMs Elizabeth and Victoria were dismantled. As the complexity of the Farringdon construction site developed over months and years it became clear that the usual practice of removing TBMs by lifting the component parts back to street level would not be practical. Instead, the drive Y TBMs were driven into the eastern end of Farringdon station, where platform enlargements had been constructed, and were then stripped and dismantled underground, with back-up and shield contents wheeled back along the newly dug tunnels for removal via the shaft at Stepney Green. The cutterheads were cut into small pieces and taken out via the shaft at Farringdon while the front 'cans' of each machine were left underground. Elizabeth line trains now pass by them.

Farringdon constraints also dictated the fate of Phyllis and Ada, the TBMs used for the western drives from Royal Oak to Farringdon. These were dismantled and their trailer systems removed from the tunnel via the Fisher Street shaft, leaving the front can and cutterhead concreted in place 30 metres below ground at Farringdon. (The drive X TBMs had been driven on tight curves off to the side at the ends of the Farringdon platform tunnel pilot lengths to veer into the surrounding earth). Getting the front sections out was considered but would have involved constructing massive chambers. One option was to build drive-on lifts but assembling a sufficiently capable crane on top of the historic underground arches around Farringdon was viewed as problematic.

A further concern was the delay – estimated at around four months – that would have been involved in retrieving the entire machines. Getting each shield out would also involve cutting them up with acetylene torches which would not only be time consuming but involve retaining the heavy duty ventilation hoses that ran all the way from the Royal Oak Portal. That in turn could have put work to break into the running tunnels from platforms at Bond Street and Tottenham Court Road on hold as well as delaying progress at Farringdon.

Crossrail acknowledged that the TBM reception arrangements at Farringdon were not ideal[7] – stripping machines backwards is time consuming compared with the usual reception pit arrangement where TBM components can be lifted out directly. In addition, the significant amount of underground cutting and burning required during the dismantling process added to the health and safety hazards of working underground.

For Mary, Sophia, Jessica and Ellie – the machines used for the shorter tunnel drives in east London and arriving above ground with fewer space constraints, the decommissioning process was far simpler. These machines were dismantled and returned to manufacturer Herrenknecht with parts recycled for use on future tunnelling projects. In 2023 it emerged that components from Ellie would feature in the High Speed 2 tunnel boring machine used to construct a logistics tunnel (delivering more than 8,000 tunnel segment rings and removing excavated material) linking North Acton with Old Oak Common HS2 station.

---

[7] Institution of Civil Engineers, Crossrail project: machine-driven tunnels on the Elizabeth line, London, May 2017, Crossrail Learning Legacy, Machine driven tunnels on the Elizabeth line

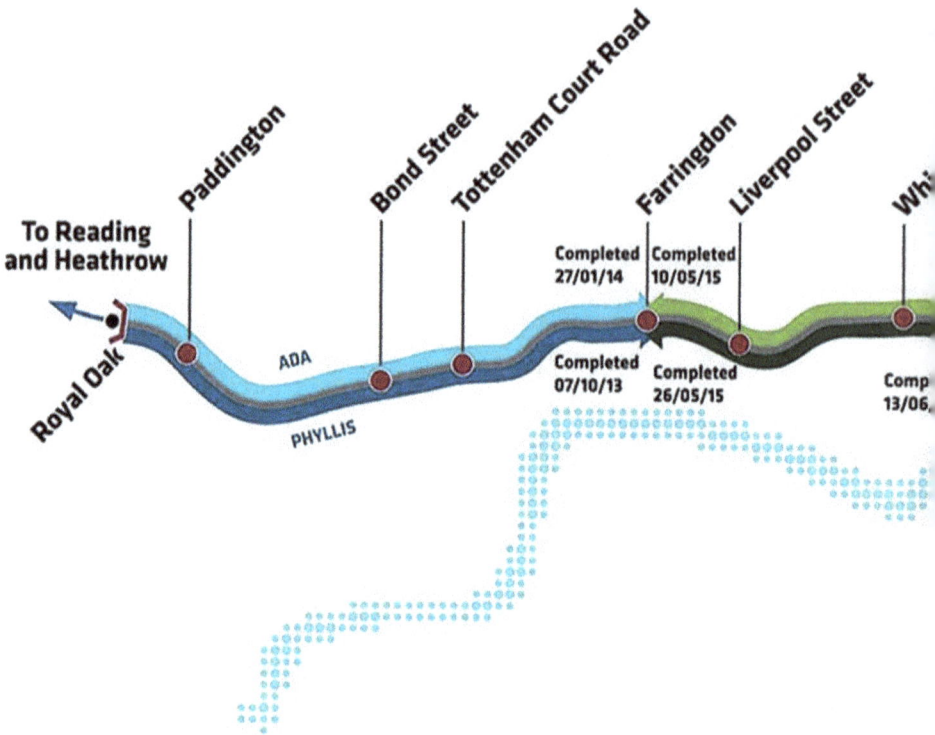

## Station tunnels, shafts and grout

Along with the main running tunnels there was a substantial amount of tunnelling required at the new central London stations and which was packaged as two contracts – station tunnels west (C410) and east (C510). These covered the construction of shafts, caverns and sprayed concrete lining works at Bond Street, Tottenham Court Road, Liverpool Street and Whitechapel.

The tunnelling sequence evolved: originally station tunnelling would have been completed before boring the running tunnels. But at Bond Street and Tottenham Court Road a decision was taken not to start the station tunnels work until the running tunnels had been completed. This made access for station tunnels work easier and cheaper but with the knock-on effect of delaying the programme.

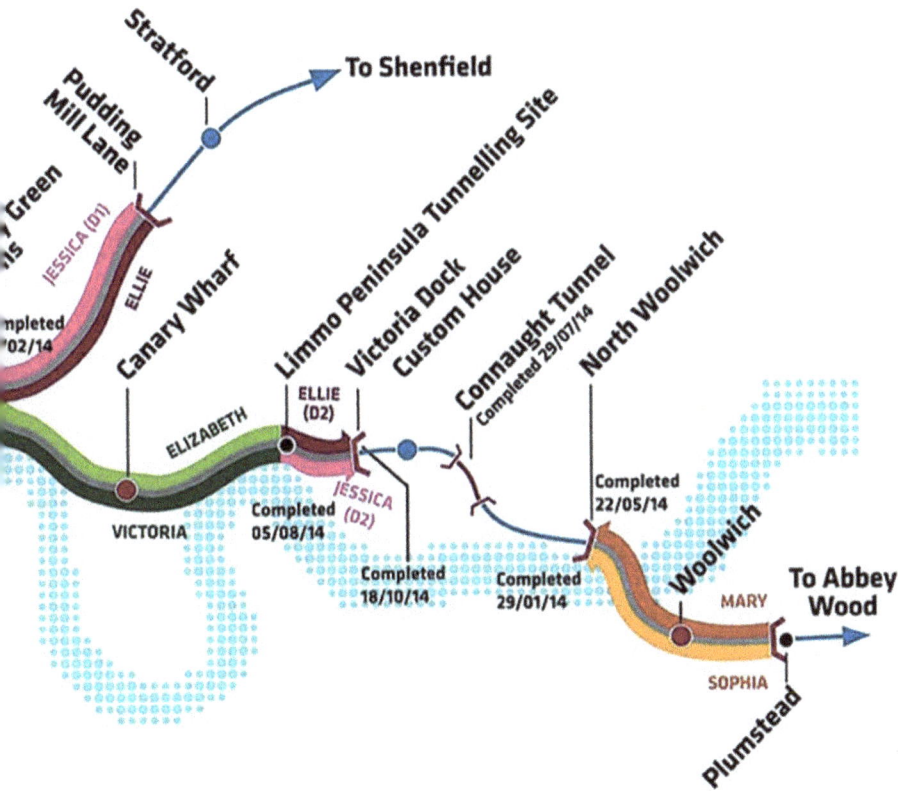

For the two eastern stations the strategy of building the station tunnels first via shafts from ground level remained. This was because the delay that would have been incurred by waiting for the longer eastern running tunnels to be bored before starting station tunnelling was deemed long enough to outweigh the advantages of easier access.

That meant that for Liverpool Street and Whitechapel, where station tunnelling would start before the eastern running tunnels were completed, a separate contractual team – Alpine BeMo, Balfour Beatty, Morgan Sindall and Vinci Construction – was brought on board.

While tunnel boring machines and precast concrete segments were used to construct the train running tunnels for the Elizabeth line, most of the underground passageways and caverns for Crossrail stations were formed using mining techniques to excavate the ground. These were then sprayed with shotcrete, a method of applying concrete projected at high velocity primarily on to a vertical or overhead surfaces[8] to form the tunnel linings.[9] Those of us following the development of the project quickly became familiar with the term 'sprayed concrete lining' and clearly this technique (think gantry mounted hoses that spray concrete instead of water) provided considerable flexibility for the builders, allowing underground 'free form' spaces to be created quickly without having to procure custom-made, precast tunnel linings of all shapes and sizes.

Sprayed concrete lining would have a major role during Crossrail construction with the technique used to build 12km of station platform tunnels, passages, access and grout tunnels along the route, 7.5km of which would become permanent features of the new railway.

Crossrail's first two sprayed concrete lining tunnels were constructed under Finsbury Circus in the City of London as part of early work for the new Liverpool Street station. The 4.5 metre diameter tunnels were built from the main access shaft within the Finsbury Circus worksite but despite their size were only temporary structures used for compensation grouting.

Wherever tunnels were built Crossrail needed to ensure stable ground conditions and this meant filling any fissures or underground voids that could result from TBM activity. Therefore, to a greater or lesser extent, depending on the geotechnical conditions of a particular site, contractors would inject grout into the ground that would fill any underground cracks and other spaces. At Finsbury Circus small diameter 'Tubes-a-Manchette' were drilled and installed from the temporary tunnels, allowing the contractor to inject grout

---

[8] American Concrete Institute, https://www.concrete.org
[9] Colin Niccolls, Crossrail Whitechapel and Liverpool Street platform tunnels project manager, Transport Briefing, 4 May 2012

into the ground. These would stabilise the ground and limit surface settlement.

Indeed, grouting was one of those behind the scenes essentials with Crossrail taking possession of sites across London that would allow it to inject fillers to stabilise ground for the tunnel boring machines. One of the first Crossrail grout shafts to be sunk was at the Bond Street eastern ticket hall site in Dering Yard. In September 2011 Costain/Skanska began building a 17 metre deep structure and similar shafts were subsequently constructed nearby in Tenterden Street, Haunch of Venison Yard, Davies Mews and South Molton Lane.

Spray concrete operator at work underground.
The sprayed concrete lining technique was widely used by
Crossrail to create free-form tunnels and caverns.

One of the more interesting Crossrail 'grout' stories was the decision by western tunnels contractor BFK (C300) to use the old Kingsway tram subway in Holborn, closed in 1952, to build an eight metre deep, five metre wide shaft to inject a cement-like substance into the ground to protect nearby buildings from any potential ground movement when Crossrail tunnel boring machines Phyllis and Ada came through in 2013.

Historically, the tunnel underneath Kingsway was used by trams carrying passengers from Holborn to Waterloo Bridge, providing the only link between the north and south London tram networks. After opening in 1906, serving two subterranean stations at Holborn and Aldwych, the tunnel was enlarged in 1929 to accommodate double deck trams.

Part of the southern end of the subway was converted into the Strand Underpass, which opened to road vehicles in 1964, but for the past 70 years the remainder has seen only occasional use – to store street furniture, as a film set and as the venue for an art exhibition. The structure is, however, grade II listed which meant BFK, once finished with the grout shaft, had to follow a precise schedule of restoration which included the reinstatement of tram rails and cobblestones.

## Stepney Green caverns

Although sprayed concrete lining has been used widely at the Elizabeth line underground stations it was at Stepney Green in east London where the technique had its most dramatic application.

Stepney is not a station but it's crucial to Crossrail because it is the point where the route splits into two (towards Shenfield or Abbey Wood) or where two routes merge into one (towards Paddington), depending on your direction of travel.

On most Elizabeth line maps this arrangement resembles a Y lying on its side, with the branches pointing to the right. But, because the Elizabeth line is made up of twin tunnels, one for each direction, where the branches meet the alignment is actually two Ys next to each other. You will appreciate that this results in one branch of each Y crossing over the other and so, to accommodate Elizabeth line train movements, a grade separated junction at Stepney means trains from the Shenfield direction pass over trains from Paddington bound for Abbey Wood.

Tunnel engineers were required to create additional space underground to provide room for the junctions and clearance for 11-carriage trains taking or arriving from either branch where two routes become one, or vice versa. The result is that each Y has a vast turnout cavern which provides the clearances required for trains to curve round in either direction.

Two smaller caverns have been built along the 'stalk' of each Y where there is a connection between the two tunnels and which during construction could be reached through a 34.5 metre deep shaft, allowing contractors access from the surface near Stepney Green Farm. This six-storey shaft was the starting point for work to hollow out the two large turnout and two much smaller caverns using mini-diggers.[10]

The Stepney Green caverns are among the largest mined underground caverns ever built in Europe: the eastbound cavern is 50 metres long, 13.4 metres wide and 16.6 metres high at its widest point. Sprayed concrete has been vital to create the Stepney caverns: as part of the C305 eastern tunnels contract the Dragados/Sisk team had to excavate 7,500 cubic metres of material and apply 2,500 cubic metres of shotcrete to the walls of the eastern cavern. The use of sprayed concrete lining allowed teams to excavate a variety of tunnel diameters and layouts including 90 degree corners – something that would have been impossible using a tunnel boring machine. The two main caverns at Stepney were completed by October 2013.

## Tunnel shafts

The length of Elizabeth line tunnels, with more than 14km between some portals, necessitated the building of shafts at locations along the main running tunnels. Although these were used during construction

---

[10] The shaft remains in use for ventilation purposes and to allow emergency access

their function now is to provide tunnel ventilation and emergency access to and from tunnels, including access for the London Fire Brigade. Following construction 'head houses' were built above the shafts with some interesting architectural finishes applied.

Some of the locations of shafts along the Elizabeth line route have already been mentioned. Outside the station developments standalone shafts were built at:

- Fisher Street (near Holborn)
- Stepney Green
- Eleanor Street (north east branch)
- Mile End Park (north east branch)
- Limmo Peninsular

There are also station shafts at Durward Street (Whitechapel) and Blomfield Street (Liverpool Street).

Construction of the 15 metre diameter Fisher Street shaft was completed at the beginning of December 2013 to support the installation of a crossover between the train running tunnels below Red Lion Square at Holborn. Although not quite on the scale of the Stepney Green turnouts it involved mining some of the largest underground caverns on the Crossrail project so that Elizabeth line trains could switch from one track to another.

Altogether 7,000 cubic metres of material was excavated as the shaft was dug and lined. Crossrail and contractors worked with UK Power Networks to relocate electricity substation equipment, a requirement before tunnel boring machines Phyllis and Ada could pass under Holborn on their journeys towards Farringdon.

On 7 March 2014 a worker at the Fisher Street shaft site in Holborn died in the early hours of the morning. The accident occurred about 10 metres underground during work to build crossover tunnels. A member of the construction team was spraying concrete on to excavated ground when concrete from the roof of the tunnel fell and hit him.

An investigation by the Health and Safety Executive concluded that Renè Tkáčik, 44, was killed when nearly a tonne of wet concrete was poured on him. Contractor BFK was fined just over £1 million for this fatality and two serious injuries which occurred on the same site. The HSE's Head of Operations said "had simple measures... been taken, all three incidents could have been prevented".[11]

Sadly this was not the only fatality to occur during construction of Crossrail. Although efforts were made to move building supplies and excavated material by rail and barge, as will be described shortly, most of the transport to and from work sites relied on lorries. Organising vehicle logistics through busy central London streets posed an array of challenges not least of which was how best to ensure safety. Crossrail Ltd insisted that heavy goods vehicles working on the programme be fitted with additional safety features[12] to protect cyclists and many of these initiatives have gone on to become standard practice in the construction industry.

These efforts proved to be not enough. During Crossrail construction there were four fatal collisions involving lorries working for project sub-contractors. Three of these involved cyclists with a pedestrian killed in the fourth incident. Transport for London has been asked by the Mayor of London to explore the possibility of creating a permanent memorial to the four victims Maria Karsa, Brian Holt, Ted Wood and Claire Hitier-Abadie.[13]

As well as the five Crossrail-construction related fatalities Transport for London says 321 'Level 2' incidents resulting in injury occurred

---

[11] BBC News, Crossrail worker death: Firms fined £1m for safety breaches, 28 July 2017
[12] For example Fresnel lenses (mirrors which show what is in a driver's blind spot), side scan equipment which results in an audible beep in the driver's cab when a cyclist is on the left inside space, and under-run guards.
[13] Mayor's Question Time, 19 May 2022

from April 2013 to 8 March 2019. Level 2 is described as an injury which leads to time off work as a consequence of the incident.[14]

## Digging the dirt

A tunnel, by definition, is a vast void underground. How ever you create it, the implication is that whatever was there before the tunnel is no longer there. Where does this 'non-tunnel' go?

To deliver 42km of tunnels Crossrail was forced to mobilise a vast logistics operation to remove the excavated material out of central London and find a permanent home for it. For the western tunnels alone 24km of conveyor belt was used to bring the spoil out of the tunnels – but once out where could it be put?

Helping Crossrail with this challenge was an initiative by the Royal Society for the Protection of Birds which developed plans for a 1,500 acre wetlands nature reserve at Wallasea Island near Southend-on-Sea. To create two and a half square miles of tidal wildlife habitat it needed around four million tonnes of clay, chalk, sand and gravel – just the kind of material that would be dug out during Crossrail tunnelling.

This raised the question of how to get the excavated material from central London to its destination out east. The strategy devised: material from the eastern running tunnels would be loaded on to ships directly from the tunnel conveyor surfacing at Instone Wharf on the Bow Creek while spoil from shafts and stations would be brought by road to Crossrail's Docklands transfer site at the Barking Riverside jetty.

For the material dug out of the western tunnels Crossrail signed a deal with aggregates firm Lafarge to use the old cement works site at Northfleet as a loading facility where material excavated from the tunnels could be lifted out of freight rail wagons and dumped on to barges (operated by a joint venture of BAM Nuttall and Dutch-owned marine contractor Van Oord) that would transport the spoil

---

[14] TfL Freedom of Information request response, 8 March 2019

onwards to Wallasea. Why Northfleet? Well, the site was one of only a few in and around London with access to both the national rail network and deep water wharfage.

But getting the spoil to the waterside would require a mini-rail project in its own right, part predicated on Lafarge's plans at the time for a new aggregates import facility at Northfleet. A rail freight link would support this but during construction of the nearby Channel Tunnel Rail Link what had become known as the 'merry-go-round' freight loop at the cement works site had had to be dismantled.

This would now be reinstated through a £13.5 million project to provide a 2.25km link between the rail network and the deep water wharf through a tunnel in the chalk cliff that surrounds the Lafarge site.[15] Compensation was available from the CTRL programme for the loss of the old merry-go-round while Crossrail agreed to chip in to fund the Northfleet work. Lafarge oversaw delivery with Balfour Beatty Rail awarded a design and build contract and Chunnel Group taking on enabling works for new sidings.

The line through Church Path Pit (between the North Kent line and Lafarge site) had already been built in 2009 as part of the final ancillary works for the Channel Tunnel Rail Link. On 23 January 2012 the Lafarge-led project team connected this to the national rail network, close to Northfleet station. Two lines merge before running under the North Kent line, with High Speed 1 and Ebbsfleet station on the right, and then curving round through the Church Path chalk pit and entering the tunnel.

The track emerges at the former cement works site where it branches into three sidings. Crossrail tunnel spoil trains used one of these, Lafarge's cement import terminal another with the central road functioning as a shared turnaround.

---

[15] The Lafarge name has mostly disappeared in the UK following a series of takeovers in the construction and aggregates sector

In February 2012 western tunnels contractor BFK appointed GB Railfreight to transport tunnel spoil to Northfleet from Westbourne Park where sidings had been built to allow material coming out of the Royal Oak Portal on conveyor belts to be loaded directly into wagons. GB Railfreight allocated two Class 66 locomotives as well as two sets of 27 JNA wagons for the job – each of the 54 wagons could carry up to 76.9 tonnes of material excavated to build the Elizabeth line tunnels.

A test run took place on 27 April with the first Westbourne Park to Northfleet train carrying tunnel spoil operating on 18 May. After that up to five freight trains a day ran Monday to Saturday with up to four trains on Sundays. During the first 15 months of tunnelling more than 860 train loads of spoil from the western tunnels were transported from Royal Oak to Northfleet.

To get a freight train from west to east London (without Crossrail) involves traversing many of the capital's lesser know rail routes. From the Royal Oak Portal trains headed west from Westbourne Park along the Great Western Main Line, through Ladbroke Grove to Old Oak Common West, then via Park Royal to Greenford, across the Greenford East Curve and up the Greenford line to West Ealing. Locomotives with wagons then travelled via Acton West to run up to Acton Bank, along the North London line at Acton Wells, passed through Willesden South West Sidings and on to the West London line, through Kensington Olympia and Latchmere Junction, Clapham High Street, Lewisham and onwards to Kent.

To allow for the increased tonnages that would be carried by the tunnel spoil trains the Greenford Loop required work on some of its embankments and track. Work was also undertaken at Drayton Green Tunnel, South Greenford, Brent South Viaduct and Greenford East Curve with track improvements required at Alperton Lane, Brent North Viaduct and South Greenford station.

## Connaught Tunnel

Any builder will tell you that it's easier to start from scratch than patch up an old building which has its own particular challenges, some known and some which will only become evident once you are well and truly committed. And so it was with the Connaught Tunnel, linking Canary Wharf to Woolwich with a tunnel connecting the Royal Victoria and Albert Docks – nice for Crossrail to have part of the tunnelled route already built but what would be the best way to restore it ready to function for decades to come?

From the north the rail alignment descends below distinctive brick arches – although earmarked for demolition these were ultimately retained and repaired. Prior to Crossrail upgrade work the route then entered a single, cut and cover tunnel wide enough for two tracks. A few metres in was a ventilation shaft to the surface above and then the railway split into double-barrelled, binocular tunnels about 100 metres long.

These existing tunnels were too small for Elizabeth line trains and the overhead catenary that would be required. South of the dock the twin tunnels joined a single tunnel by another ventilation shaft and, via more brick arches, the alignment returned to surface level.

Upgrading the binocular section for Crossrail presented difficulties. In 1935 when the Royal Victoria dock above was lowered to accommodate larger ships, over-enthusiastic dredgers managed to expose the Connaught Tunnel crowns. Amid fears that the whole thing could flood at any second a decision was taken to reinforce the twin tunnels by fixing metal straps inside. Crossrail planned to take these out and then widen the tunnels further to provide the space required for the Elizabeth line service.

An initial proposal to remove the steel tunnel lining straps, fill the existing tunnel with concrete foam and then rebore it to create the larger profile required for Crossrail was dropped amid concerns about the structural integrity of the tunnel. Instead, part of the dock above was drained with workers then digging down through the dock bed to access the tunnel roof. From there they could enlarge the tunnel profile and cast a 1,000 cubic metre concrete slab to sit above the tunnels and effectively thicken the dockbed. This would be the first time the Connaught Tunnel had been exposed from above ground since its cut and cover construction in the 1870s.

Connaught Tunnel contractor Vinci Construction (C315) certainly had an interesting commission. One of the challenges of working on a 140 year old tunnel was that construction blueprints were hard to come by – of those that had been unearthed no one was sure how accurate the plans were. One drawing showed the existence of three service tunnels but other historical records made no reference to these. Part of Vinci's brief was to carry out exploratory work to ensure no-one drilled into any void that might cause flooding or other problems.

The contractor was also given the job of demolishing the overground Silvertown station, left to decay since the North London line North Woolwich branch closed in 2006. Despite talk of building a new station nearby no provision for this was included in the Crossrail Act.

Following installation of a cofferdam to isolate the water in the dock above the tunnel, draining of this part of the dock began in early 2013, timed to begin after the London Boat Show (the London ExCeL exhibition centre is nearby) and only given the green light following survey work to check for potential unexploded ordnance from World War II in the tunnel's western approach. Thirteen million litres of water were drained and then the dock was refilled towards the end of the year in time to be used by vessels attending a military trade show at ExCel.

## Beyond tunnelling

So having dug a tunnel, what happens next? The completion of each section of Crossrail tunnel would mark the start of a complex programme to install different systems, many of which are detailed later in this book. But the most fundamental system required for a railway tunnel is the permanent way, the track on which trains run.

For the Elizabeth line the standard sequence was for the tunnel builders to pour the first stage concrete – a tunnel floor with a kerb that provided a guide for the rails that would slot inside. Welded rails were then brought into the tunnel and laid on the ground. Systems contractor ATC (Alstom, TSO, Costain) used a vast multi-purpose gantry which slotted track slab into position and then lifted up the rails, threading them into position on top of sleepers.

The Elizabeth line has five types of track slab. With a total length of 41.2km, standard track slab accounts for around 80% of the central section of Crossrail. Direct fixed track was chosen for use in the Victorian engineered Connaught Tunnel beneath the Royal Docks, allowing engineers to create a flat surface on top of significant undulations in the ground and to work within height restrictions.

High attenuation sleepers – similar to standard slab – were used in a few small areas to reduce noise and vibration with floating track slab used to minimise noise and vibration along parts of the route. The track slab floats on a combination of elastomet rubber bearings and heavy duty springs and a light version was used between Tottenham Court Road and Bond Street under Soho with a heavier version fitted where the Elizabeth line passes close to the Barbican concert venue.

With rails correctly aligned the slab then needed to be fixed in place – a job for ATC's 465 metre long concreting train. Based at Plumstead, the concreting train headed west with all the supplies required on board to produce a suitable mix that could then be hosed into position by the systems team on the ground. By this point the railway was recognisable as such.

With the concreting train operating from Plumstead, track-laying began along Crossrail's eastern stretches, including through the Connaught and Thames tunnels, typically covering 170-180 metres a day, working primarily at night. A second permanent rail laying and concreting operation began along the west of the route in March 2016 but without the use of a concreting train.

After track had been concreted in place the drilling rig followed. Built by Rowa Tunnelling Logistics this impressive machine fulfilled the need to drill a lot of holes – approximately 250,000 – in order to attach the brackets for 25kV overhead line electrification, emergency escape walkways, signalling cables and more to the concrete tunnel segments.

To achieve this locations for the holes were programmed into the drilling rig with positions calculated to avoid the joins between concrete segments and other inappropriate locations. Lasers were used to match up the physical tunnel layout with the drilling rig's blueprint and the machine was then set to work. This provides an example of how the dirty engineering of the past has been transformed into a clean and precision-based procedure: each drill bit on the machine had vacuum abstraction to Hoover the concrete dust away. Using a drilling rig also avoided vibration-induced workers' injuries such as 'white finger'.

## TUCA and training

The absence of a UK tunnelling pipeline, an ongoing stream of projects with an established talent pool, meant the decision to commit to 21km of double Crossrail tunnels would create a sudden demand for people with specific tunnelling skills. At the peak of construction Crossrail required around 14,000 workers – 3,500 of whom needed training in underground construction techniques – and it was against this background that the UK's new Tunnelling and Underground Construction Academy (TUCA) was built, a distinctive red hangar, in Aldersbrook near Ilford in east London.

TUCA would not be just for Crossrail – it now fits into Transport for London's portfolio – but Crossrail was the catalyst for its inception and promised up to £7.5 million for the training centre with the government committing a further £5 million.

When it opened in summer 2011 TUCA began to provide training on the key skills required to work in tunnel excavation and underground construction. The only other tunnelling training facility in Europe at the time of opening was located in Switzerland; this specialised in hard rock tunnelling whereas TUCA focused on teaching the soft ground tunnelling skills more sought after in the UK.

Set over two levels, the academy was designed to replicate the key areas of a fully-automated tunnelling project. It had a simulated tunnel boring machine environment supported by a TBM backup area with a loco and narrow gauge railway to the rear. A separate hangar could be used for sprayed concrete training. There was also an underground construction workshop, four teaching rooms, a test centre, refectory and a learning resource centre.

All those working underground on Crossrail were required to obtain a Tunnel Safety Card before they could start work. This is a specially developed health and safety test which has become a unit of the nationally recognised Construction Skills Certification Scheme Health and Safety Test.

The academy's initial training programme was focused on delivering suitably skilled and safe workers who were qualified to work underground. By September 2011, besides the Tunnel Safety Card, training at TUCA included the NVQ Level 2 in Tunnelling Operations and NVQ Level 3 in Tunnelling Operations Supervisory Skills. Training was delivered by a specialist skills provider with the National Construction College, the training arm of the Construction Industry Training Board (CITB), awarded a five year contract in 2011 to develop a suitable curriculum and deliver a full range of accredited and bespoke skills programmes.

In 2017 responsibility for TUCA was transferred to TfL which awarded Prospects College of Advanced Technology the contract for delivering training. However, financial difficulties faced by the Essex based provider led it to announce a merger with South Essex College in 2018. At the time of writing the TUCA building was owned and managed by TfL with subsidiary Rail for London Infrastructure installed as 'tenant'.

During construction of the Elizabeth line most students attending the academy came from contractors' existing workforces but Crossrail partnered with Jobcentre Plus to ensure that local people along the Crossrail routes were informed of training and employment opportunities on the programme. All contractors were required to advertise any new opportunities with Jobcentre Plus 48 hours before they were advertised elsewhere.

Apprenticeships provided another route to a Crossrail job. All main works contractors were obliged to deliver one apprentice per £3 million contract spend. More than 1,000 apprenticeships were created during the course of the Elizabeth line programme by Crossrail Ltd, Network Rail and Bombardier Transportation.[16]

## Core storage

Understanding ground conditions is a prerequisite for any tunnelling enterprise and has required Crossrail to collect thousands of earth and rock samples. To store this burgeoning geological collection Crossrail turned to the UK's largest salt mine in Winsford, Cheshire, which provides a climate controlled storage facility.

Samples have been stored since 1992, 17 years before Crossrail construction began in 2009. During development and construction of the Elizabeth line soil and rock cores were extracted from more than 1,400 boreholes along the route at depths of up to 70 metres. After being removed from the ground soil and rock cores lose moisture and slowly deteriorate but the consistent humidity level and ambient temperature of 15 degrees celsius at Winsford DeepStore prevents deterioration without the use of an energy-intensive, temperature-controlled warehouse.

The soil and rock cores at DeepStore include London clay, sands and clays of the Lambeth Group, Thanet Sands and chalk. Together, they tell the story of how geological activity has shaped the earth on which London sits over the past 35 million years. The oldest Crossrail core is said to be 80 million years old.

Once a sample had been collected it was carefully examined, tested and catalogued to ensure data was available to inform the designs

---

[16] TfL press release, One year to go until the beginning of Elizabeth line, 19 December 2017

of all Crossrail tunnels, portals and shafts. This could then be used to ensure engineers excavated tunnels and stations safely with minimal subsidence risk to the buildings and infrastructure above. The samples may also inform the development of future railway tunnelling projects such as Crossrail 2.

## Tunnelling complete

By the middle of 2015 Crossrail had its train running tunnels. Ten tunnels built by eight tunnel boring machines had been connected to form an additional 26 miles (42km) of underground railway made up of more than 200,000 concrete segments. The new infrastructure for the Elizabeth line passed over, under or next to 31 London Underground locations including the Bakerloo, Victoria, Central and Northern lines with minimal ground movement recorded and no adverse impacts on LU infrastructure.

Dedicated tunnel gangs, using the extraordinary Herrenknecht tunnelling machines, were essential to this achievement, and to them belongs much of the credit for creating structures that now form the fabric of London's expanded rail network. But, as we have seen, Crossrail tunnelling involved a diverse range of other projects, all interlinked and requiring ingenuity at many levels.

While tunnelling progressed the carving out of cathedral-like structures that would form the Elizabeth line's new stations was also underway. Much of the engineering know-how from building the running tunnels would be important here but in addition there would be a focus on finish – how the stations look and feel. After all, in the tunnels trains separate and insulate passengers from raw concrete. Could the station construction teams turn vast building sites into facilities people would actually want to use?

# 5. THE STATIONS

Construction of the Elizabeth line has transformed London's station facilities.

The railway has 41 stations in total. Many of those on the outlying surface routes have been significantly upgraded; details can be found in chapter 6. But on the central section of the route the work is on a different scale; billions of pounds have been invested creating new or completely rebuilt stations.

Discussions of the Elizabeth line 'entity' can be deceptive because this label masks the fact that many of the new station schemes were mega-projects in their own right. Removed from the Crossrail portfolio and presented as a standalone scheme many would find few rivals in terms of scale or ambition anywhere in the UK.

As an observer I would certainly be impressed by any project to build a station with more than 50 kilometres of communications cabling, 200 CCTV cameras, 66 information displays, 200 radio antennas, 750 loudspeakers and 50 help points. Yet this was the baseline spec for each of the new central London Elizabeth line stations.

Impressive as this kit list is, only by visiting the new stations can you appreciate just how much has been achieved and how different they are from stations previously built in London. Put simply, the new underground stations are vast. While keeping costs down is always a priority for infrastructure schemes Crossrail has managed to learn lessons from projects such as the Victoria line by building not only space for today but room for the system to expand and accommodate future demand.

The designers have questioned many of the assumptions of building underground railways to come up with innovations in lighting, ventilation and wayfinding that lay down a new benchmark for metro systems. They even went so far as to build a mock-up platform tunnel (which I got to visit in 2011) at Vinci Construction's Technology Centre in Leighton Buzzard, Bedfordshire.

The mock-up helped Crossrail and contractors understand how the designs for the new platforms would translate to real life and was used to inform design decisions, including signage and platform acoustics. Having the mock up persuaded designers to relocate next train information from drop down displays spaced at intervals along platforms to screens directly above platform edge doors.

Perhaps the sharpest contrast between London's old and new underground railways is where platforms and subways connect and intersect. The use of sprayed concrete lining at Elizabeth line stations has led to curved junctions and spaces that are larger than existing LU assets. Glass-fibre reinforced concrete has been used to clad the structural tunnel lining leaving smooth, sweeping, curved edges designed to ease passengers' journeys underground.

The new Crossrail stations have lots in common but they are also defiantly different from each other. Four have ticket halls at each end, maximising their catchment areas. Elizabeth line station platforms have a carefully curated consistency but as passengers make their way to the surface, through the ticket halls and into the urban realm outside the station, this underground consistency gives way to individual designs reflecting local character and context.

Realising the visions of the designers has not been easy. Enabling works for most stations began in 2009-10 with main construction underway at all sites by the middle of 2013. Costs rose, delivery dates slipped, Covid arrived... problems at some of the stations have been widely reported. In some ways the building of the tunnels and shafts was the easy part. After fit-out half a million individual railway assets – from smoke alarms to CCTV cameras – had to be tested and certified for use before the new stations could open to passengers.

Read on to find out more about the nine central Elizabeth line stations and how they were built. Details of the Abbey Wood redevelopment, which was contracted by Network Rail, can be found in chapter 6.

'Under The Glacier' artwork by Darren Almond on
display above escalators at Bond Street station.

# BOND STREET

| Role | Responsibility |
| --- | --- |
| Architect | John McAslan and Partners |
| C132 structural design | WSP |
| C207 demolition works | McGee Group |
| C223 enabling works | Select Plant Hire |
| C240 utility diversions | McNicholas |
| C411 advance station works | Costain/Skanska |
| C412 main works | Costain/Skanska |
| Additional design (originally sub-contracted to C412) | Arup/Hawkins Brown |
| Additional final fit-out | Engie |

The Elizabeth line does not have an Oxford Street station but the new railway runs parallel to London's famous shopping thoroughfare. The double ticket hall design of Tottenham Court Road and Bond Street Crossrail stations provides four points from which to access this part of the West End.

The western Bond Street ticket hall has been built on Davies Street with red sandstone precast columns forming a colonnade to frame the entrance. It connects to the existing Bond Street London Underground station via a subway running under Oxford Street and complements a £300 million project to expand the Tube station.

At the other end of the Elizabeth line platforms the station's second ticket hall is accessed from Hanover Square, again just south of Oxford Street and a short distance from Regent's Street and Oxford Circus. Bond Street's eastern ticket hall also uses a colonnade design, this time featuring pale Portland stone. Here you will find the longest escalator on the Elizabeth line – at 60 metres this is now the second longest on the Transport for London network with only the escalators at Angel Underground station coming in longer by a mere one metre. While we're on the topic, there are a total of 56 lifts and 81 escalators on the central section of the Elizabeth line. The shortest Crossrail escalators can be found at the canopy entrance at Liverpool Street, connecting street level with the Broadgate ticket hall.

## Building Bond Street

At Davies Street the western ticket hall structure was formed by excavating a 25 metre deep box with five intermediate levels directly above the London Jubilee line. Tension piles were installed to support the older underground railway.

In March 2013 tunnel boring machine Phyllis passed just to the south of the box followed by Ada;[1] her route went directly through the walls of the then-unexcavated box. Both machines then continued to the Bond Street eastern ticket hall at Hanover Square. Here two 25 metre deep circular shafts were constructed to provide

---

[1] Crossrail's TBM Phyllis Reaches Bond Street Station, Tunnel Business Magazine, 28 March 2013

early access down to track level in order to support the western tunnelling drive. Ada broke through into one of these shafts where the cutterhead underwent maintenance before the TBM continued her journey east. A small reception chamber was built to allow access to the cutterhead of her sister machine. As with the western ticket hall there are five intermediate levels at Hanover Square.

Following excavation work structures for the new ticket halls were built above ground including chimney-like ventilation structures to allow air to circulate to platform and sub-surface levels. These would be a temporary aesthetic with oversite property developments soon enveloping them with modern office accommodation. At Hanover Square Great Portland Estates was responsible for delivering a 300,000 square foot mixed use development of retail, commercial and residential space. The temporary exposure of the vent shafts prompted the government's architectural watchdog to say that it was "concerned by the choice of perforated, anodised, aluminium panels as the cladding material for the vent shaft," adding 'we do not consider this metal cladding to be acceptable, even as a temporary treatment."[2]

In September 2015 the expanded Bond Street London Underground station was connected to the new Bond Street Elizabeth line station with the completion of the pedestrian subway below Oxford Street. By this point the Davies Street ticket hall was structurally complete with excavations continuing at the eastern ticket hall. All the station's connecting passageways had been dug and, by the end of the year, construction of the 250 metre long platforms was nearly finished. The upgrade of Bond Street London Underground station – contracted to a Costain/Laing O'Rourke joint venture and including a new Marylebone Lane entrance, additional escalators to the Jubilee line and expanded ticket hall – completed in November 2017.

## Last to finish

Despite a promising start building and fitting out Bond Street did not go to plan or adhere to the programme schedule. As detailed in chapter 11 it would take another five years for the station to be ready

---

[2] Bond Street East Crossrail Station Review, CABE (later merged with the Design Council), February 2011

for passengers – it remained shut when the rest of the Elizabeth line opened in May 2022, only finally opening to passengers in October.

What happened? Unforeseen ground conditions, delays to station works while tunnel boring machine back up systems occupied platform areas and restrictions on working practices imposed by Covid lockdowns all contributed to construction and fit-out delays.

Perhaps it didn't help that Bond Street was the final main central London station contract to be let (in early 2013) as Crossrail Ltd staggered procurement in an attempt to avoid overwhelming the construction industry with a glut of station mega-contracts.

But ultimately much of the delay and cost can be attributed to muddled interfaces between contractors and a system that encouraged companies to make compensation claims and counter-claims against each other. In June 2020, as delays and costs mounted, Transport for London and the main contractor at Bond Street, Costain/Skanska, parted company, enabling TfL to work directly with Arup and Hawkins. Although this created challenges, requiring TfL to co-ordinate the work of many smaller suppliers, the Greater London Assembly estimated bringing this work in house reduced the final cost of completing the station by up to £76 million.[3]

Now the station is open there's lots to like. The columns at both ticket halls blur the line between interior and exterior spaces and bronze detailing abounds in ventilation grilles and sound absorbing ceilings. Outside the Portland stone faced eastern ticket hall at Hanover Square is the Medici Courtyard, said to be the first public courtyard in Mayfair to open in more than a century, developed as part of a Crossrail public realm scheme.

Like the other major new Elizabeth line stations Bond Street comes complete with specially commissioned artwork, in this case three

---

[3] GLA Procurement, GLA Oversight Committee, July 2022. It cost £19 million for Crossrail to exit the contract with Costain/Skanska but was forecast to cost up to £95 million to complete the station under the original contractual arrangements

abstract pieces by British artist Darren Almond. 'Horizon Line' consists of 144 hand-polished tiles, which reference daily routines, timetables and schedules, while 'Shadow Line' and 'Time Line' resemble the embossed metal nameplates fixed to early British locomotives only here they bear a selection of poetic phrases. These may be pondered by travellers passing through the station.

Platform level at Canary Wharf could only be constructed after draining the dock above.

## CANARY WHARF

| Main contractor | Canary Wharf Contractors |
|---|---|
| Piling and enabling works | Expanded (Laing O'Rourke) |
| Lead designer (geotechnical, structural and building services) | Arup |
| Operations architect and design | Tony Meadows Associates |
| Project architect | Foster + Partners |
| Glulam timber roof | Wiehag |
| Rooftop park landscape designer | Gillespies |

Resembling an island in the North Dock of West India Quay, Canary Wharf – Isle of Dogs station as it was know prior to a funding agreement being signed – is one of the largest Elizabeth line stations. The station, retail and park areas are six storeys high and at 256 metres the development is slightly longer than the height of nearby One Canada Square, one of the tallest buildings in the UK. The station development links Canary Wharf with Poplar, previously separated by the North Dock, and includes connections to the Canary Wharf Estate, via Adams Place, and the Jubilee line and Docklands Light Railway Canary Wharf stations.

From early in the Crossrail construction programme Canary Wharf was the place where one could go behind the scenes and see the build in action; the media were regularly invited to come and view what was going on. This was in part because of the unique funding arrangement for this station. Contracts for the other new stations were let by Crossrail Ltd, paid for out of its funding settlement, and contractors were therefore subject to the standard suite of confidentiality protocols which required anyone thinking of speaking to the press to go through Crossrail first.

Canary Wharf station was different. In December 2008 the government signed a deal with Canary Wharf Group, the property company which owns the central Docklands business district, to design and construct the station for a fixed price of £500.3m, of which CWG would stump up £150m. Should it cost less CWG would retain the saving and it would, in theory, foot the bill in the event of any overrun.

It certainly seemed that Canary Wharf Group, and its construction arm Canary Wharf Contractors, were more eager than other contractors to speak to the media. But then in the early stages of Crossrail construction they had more to show off – while the other new stations were stuck in a holding pattern of enabling works the new Canary Wharf station was rapidly taking shape.

The key requirement was for the Canary Wharf station box to be finished by summer 2012, ready for the two tunnel boring machines launched from Limmo Pensinsula to pass through the station en-route to Farringdon. To complicate matters the site chosen for the station was Canary Wharf's North Dock, full of water that would have to be drained before any digging could begin.

Prior to completion of the B6 concrete slab the
Canary Wharf station box still contained Thanet sand.
The TBM arrival point is marked on the wall.

When Crossrail's first main station works construction began in May 2009 we learnt a lot about piling. More than 300 piles were sunk into the dockbed to create a 1.2 metre wide cofferdam that would form the perimeter of the worksite – which was then drained to allow construction to begin.

Canary Wharf Group was particularly proud of the use of a silent hydraulic Giken piling technique used to build the cofferdam. One of the first uses of this Japanese construction method in the UK, it allowed Canary Wharf Contractors to carry out major piling work – normally a noisy process – without unduly affecting office workers overlooking the site or guests at the nearby Marriott hotel.

After draining the dock – 40 Olympic swimming pools worth plus a lot of silt and a few old tyres – further piles were sunk in the dockbed and top down construction of the station began. This involved excavating a level and then casting a concrete slab to form the floor – and ceiling of the next level. The process would be repeated until the entire station box had been excavated. At each end of the station box earth banks were formed using spoil excavated from the box to guard against the possible impact of a runaway barge striking the new station.

## Going with the grains

At Canary Wharf basement six represents the bottom of the station, this being the level of the railway, with the B6 concrete slab approximately 18 metres below the dockbed (which in turn is 10 metres below normal water level). In May 2011 I was able to visit track level, prior to completion of the full B6 slab, and got to experience what resembled an underground beach consisting of the Thanet sand which makes up the Dockland's geology at this depth. Canary Wharf Contractors had helpfully drawn a circle on the concrete wall at one end of the station box to show where one of the tunnel boring machines would break through.

To mitigate for the loss of dock capacity caused by building the station Crossrail constructed an underwater chamber, alongside nearby Adams Place, which can be used in the event of flooding. This initiative replaced earlier plans to widen the River Lea to expand the flood plain.

North Dock was refilled in 2012. In March 2012, with the station structure complete, work split into two strands. At track level Crossrail's tunnelling contractors were given access to prepare for the arrival of TBMs Elizabeth and Victoria – managing the breakthrough into the box and then moving the machines on multi-wheeled trolleys to the other end ready for the next tunnel drive. Hoardings were erected to separate the B6 track area (with the conveyor belts carrying muck back from the tunnelling faces to Limmo) from the island platform level (B5); Canary Wharf Contractors continued working here as well as on the ticket hall (B4) fit-out above.

Indeed, Crossrail handed the contractor additional work, beyond the scope of Canary Wharf Group's original funding agreement, to install fire protection and similar building systems. The thinking was that this would avoid delays later – although it subsequently proved not to be the case when it emerged that fire systems installed at Canary Wharf did not meet project-wide standards.

With North Dock refilled Canary Wharf Elizabeth line station became an island. Above the ticket hall and platform levels Canary Wharf Group created a four-storey retail development, Crossrail Place, which opened to the public in May 2015. An elevated walkway provides a covered pedestrian route to the station shopping mall entrance. The Canary Wharf station complex has eight 30 metre long escalators with distinctive yellow glass surrounds, nine 11 metre long escalators, and six lifts. As part of the Crossrail Art Programme a 16 metre long video artwork by Michal Rovner has been installed and references the architecture of London as well as addressing themes of humanity, history and time.

Topping the retail development is a 4,160 square metre roof-top garden featuring a timber lattice roof. Taking inspiration from the Eden Project in Cornwall 780 ethylene tetrafluoroethylene (ETFE) semi-transparent pillows fit into the spaces between the timber lattice roof and are kept inflated with a weak jet of air piped around the birch glue-laminated frame. Spaces have been left unfilled in parts of the lattice to allow trees to grow up and through the roof.

In November 2015 Canary Wharf became the first Crossrail station construction project to be completed with Canary Wharf Contractors officially handing over the ticket hall and platform levels to Crossrail Ltd. Note that this did not mean the station was finished – work on equipping it with communications equipment, signalling, tunnel ventilation, platform screen doors and overhead line equipment was still to come. Further fit out work would also be required, so much so that in January 2022 Canary Wharf became the last but one Crossrail station to be transferred to TfL operational control.[4]

---

[4] Transport for London press release, 25 January 2022

The modular station at Custom House provides
an interchange with the DLR.

## CUSTOM HOUSE

| Role | Responsibility |
|------|----------------|
| Architect | Allies and Morrison |
| C146 design | Atkins |
| C520 main works | Laing O'Rourke |

Crossrail's only all-new overground station, the Elizabeth line
has returned national rail services to Custom House following the
closure of the North London line to North Woolwich in 2006.[5]

Custom House was already part of the Docklands Light Railway but
the arrival of Crossrail has transformed it from a lightweight DLR

---

[5] The Elizabeth line uses part of the former North London line route
including the Connaught Tunnel

stop to a two storey concrete colonnade structure. The 200 metre station building has been slotted into a ribbon of land between the Docklands Light Railway Beckton branch and Victoria Dock Road. The first floor concourse is above the Elizabeth line island platform faces and is connected to a series of elevated walkways. This top level accommodates ticket machines, automatic barriers and an operations room which supplies power for the station and Victoria Dock portal nearby. Capping it is a lightweight, transparent roof featuring the high strength ethylene tetrafluoroethylene plastic also used at Canary Wharf.

According to the design team the soffit of the concourse floor slab has a series of petal-shaped panels which are uplit to give a feature ceiling to the western end of the platforms. Glass canopies line the southern edge of the colonnade on the westbound platform face; at 300 metres long the platforms are not entirely covered by the concourse above so free-standing platform shelters have also been provided.

To speed up and streamline construction main contractor Laing O'Rourke's approach was to manufacture the new station structure at its factory in Steetley in Worksop before transporting 825 precast concrete segments 130 miles to Custom House for reassembly with the help of a gantry crane – a large hoist trolley that shuttled up and down the station site on a pair of rails. The main station build began in May 2014 and took a year to complete.

In July 2013 Newham Council approved plans for an additional station entrance via a footbridge spanning Victoria Dock Road. In fact three pedestrian bridges have been created. The biggest, installed in December 2014, was a 34 metre, 90 tonne structure linking the Elizabeth line station concourse level with the existing system of elevated walkways connecting the DLR station with the ExCeL exhibition centre. Previously the walkway from ExCeL ended at a T-junction with the two arms providing access to DLR platforms. The new bridge continues straight on (turning the T into a +) and after crossing the twin DLR tracks and one Elizabeth line track joins the Custom House elevated concourse (turn right if heading from ExCeL).

Bridge number two takes the pedestrian route from ExCeL straight on and over Victoria Dock Road where stairs and a lift shaft were installed at the junction with Freemasons Road. This is a clever bit of regeneration engineering which gives pedestrians a traffic-free route across Victoria Dock Road and both railways, connecting Custom House and the residential area north of the railway with the Royal Docks and ExCeL to the south. The footbridge has been designed as an extension of the public realm as well as the means to access the Elizabeth line. Both these bridges opened to the public at the end of 2015. A third bridge provides direct access between Elizabeth line and DLR platforms.

Ahead of the Elizabeth line opening Transport for London upgraded Custom House DLR station, preparing it to become an interchange. Installation of two additional staircases and a mezzanine deck is said to have increased the DLR station capacity by 50%. With the rear of the station structure backing on to Victoria Dock Road the opportunity has been taken to soften the impact of the concrete and discourage graffiti with a community artwork – the Newham Trackside Wall. Designed by artist Sonia Boyce, panels featuring pictures and facts about the local area as well as personal testimonies run for 1.9km to create one of the longest commissioned artworks in the UK.

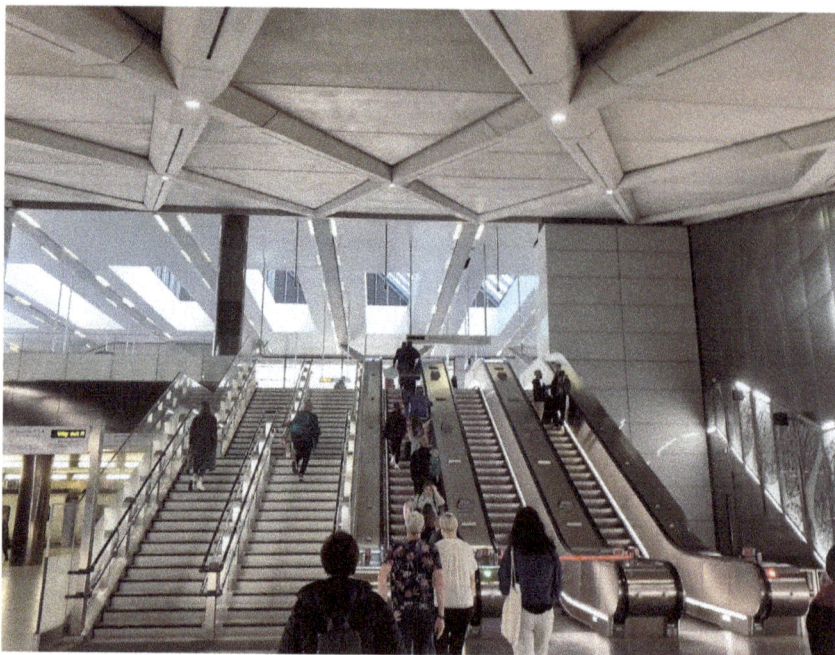

Diamond design ceiling viewed above escalators to
Farringdon west integrated ticket hall.

## FARRINGDON

| Role | Responsibility |
|---|---|
| Architect | Aedas |
| Civil and structural engineer | Aecom |
| Thameslink integrated ticket hall main works | Costain/Laing O'Rourke |
| C136 design | URS Scott Wilson |
| C209 demolition works (Farringdon east) | Keltbray |
| C430 advance station works | Laing O'Rourke/Strabag |
| C435 main works | BAM Nuttall/Ferrovial/Kier |

Farringdon (in case you were wondering it means fern-covered hill)
can make a strong claim to be the nucleus of the Elizabeth line.
During construction it was the point of arrival for tunnel boring

machines heading from both east and west. It is the point where the London clay trusted for tunnelling meets shifting Thanet sands. When construction got underway it was Crossrail's most expensive station although the estimated cost of £375 million would prove well short of the final price tag.[6]

With construction complete it remains a point where forces meet, a mega interchange with the vertical axis of Thameslink. Pre-covid forecasts anticipated that 150,000 people a day would use Farringdon, up from 60,000 a day in 2010/11.[7] Farringdon is the only station in the UK providing direct trains to Heathrow, Gatwick and Luton airports. With typical 24 train per hour frequencies in each direction on both routes, plus a train every minute or so on the London Underground Metropolitan, Circle and Hammersmith & City lines, the station can be served by approaching 140 trains per hour.

Farringdon station is designed with a ticket hall at each end of the platforms. Although passengers today experience the standard 244 metre fitted out platforms the overall length of the platform constructed in the eastbound tunnel is more than 400 metres due to the location of the ticket halls.

To the west is the pre-existing Farringdon Underground station on Cowcross Street which has now been supplemented by extensive new facilities for the Elizabeth line and Thameslink. The eastern ticket hall was built at one end of Smithfield meat market on a block along Lindsey Street between Long Lane and Charterhouse Lane – 13 buildings were demolished to make space. Access is from the corner of Lindsey Street and Long Lane with a small ticket hall – the coffered ceiling design is a nod to the brutalist architecture of the nearby Barbican centre – leading to two banks of escalators to platform level. The first run parallel with Lindsey Street and the

---

[6] Chapter 11 shows that the Farringdon main works contract cost in excess of £600 million
[7] London Underground/Office of Rail and Road

second at ninety degrees head towards the western ticket hall end of the station. Running alongside the escalators are two funicular style incline lifts; Farringdon and Liverpool Street are the only Elizabeth line stations which have these.

The eastern ticket hall site is just around the corner from Barbican station, which serves the same sub-surface London Underground lines as Farringdon plus the Great Northern route from Moorgate. Barbican station can be accessed directly from the east end of the new Elizabeth line station platforms. Farringdon also has two fire-fighting lifts for emergency use.

## Thameslink enhancements

Development of the Farringdon west site was closely linked with the Thameslink Programme, an initiative to upgrade the facilities provided since Thameslink trains started running in 1988, following the reopening of the Snow Hill Tunnel, as well as expand the Thameslink network of routes.

At Farringdon a £250 million package of station improvements was completed by Costain and Laing O'Rourke in July 2012 as part of Thameslink Programme Key Output 1. Changes included extending the Thameslink platforms southwards, severing Farringdon Junction in the process to spell the end of Thameslink services to Barbican and Moorgate (LU trains still run), extending the station roof and building new passenger walkways. A new Turnmill Street side entrance provided an additional way in and out of the station and the street-level LU station concourse would shortly afterwards be expanded to provide easier access to platforms. New lifts were installed and sunk to the depth of Crossrail platforms to avoid work that would have been disruptive (if not impossible) later on.

However, the most noticeable part of the Farringdon Thameslink upgrade was the new integrated ticket hall built opposite the London Underground entrance on the now part-pedestrianised Cowcross Street. The design concept for the 165 metre square integrated ticket hall replaced earlier plans for two separate ticket halls with a generously sized gateline wide enough to handle passengers for both Thameslink and Elizabeth line services.

The integrated ticket hall opened with the start of the national rail winter timetable on 12 December 2011, coinciding with the introduction of 12-carriage Thameslink trains. The ticket hall had a temporary partition wall inside the ticket gates on the right, as you enter, where the Elizabeth line escalators are now. Altogether the Thameslink Farringdon upgrade, which made provision for Crossrail, involved the installation of 20 staircases and 36 automatic ticket gates.

## Construction plans in flux

In addition to all the standard challenges of constructing an underground station, at Farringdon engineers also had to plan for the arrival of four tunnel boring machines (two from the west, two from the east) and volatile ground conditions caused by multiple geological faults. Extensive grouting was required at the station to compensate for ground settlement caused by tunnelling. Most of this was undertaken within the disused railway arches under Smithfield Market; existing railway tracks and station facilities were monitored hourly and if movement was detected grout was injected from one of five grout shafts to compensate for any ground settlement.

Partly as a result of the geology and concerns about drainage, an early plan was to lower a mini TBM down a shaft from the western ticket hall site which could dig a pilot tunnel for the platforms. Data subsequently showed the ground was more stable than had been thought and the costly mini TBM plan was scrapped in favour of the standard Crossrail approach of breaking through the new running tunnel segments to create space for platforms. Once the station tunnels had been constructed two reception chambers were built to receive the TBMs from the east.

Given all the complications at Farringdon it may have been helpful that both tunnelling and main station works were the responsibility of BFK; Farringdon was the only Elizabeth line station where the tunnelling, construction and fit-out was carried out by the same contractor.

In November 2015 the late Prince Philip, Duke of Edinburgh went underground to view the recently completed tunnels being waterproofed and the final concrete finish being applied. In a rare example of safety procedures not being followed the Duke was granted a royal exemption from wearing the orange high-vis trousers mandatory for all those on site.

## Delivering new ticket halls

Next to the Thameslink integrated ticket hall what is sometimes referred to as Crossrail's western ticket hall at Farringdon occupies the site of the former 1960s Cardinal House at the corner of Cowcross Street and Farringdon Road. It's more access point than ticket hall; today two banks of escalators connect the Elizabeth line to the integrated ticket hall and Thameslink platforms. During the construction programme there were three shafts on the site which changed as the Crossrail tunnelling strategy evolved.

A circular shaft was designed for lowering and launching a mini tunnel boring machine for construction of the eastbound pilot tunnel and, when this plan was abandoned, had to be modified to accommodate the passing through of components from the full-size, eastbound TBM. It was then used to remove excavated material from the platform enlargement and to complete the fit-out. Two rectangular shafts were also created – the smaller, shallower one as a starting point to excavate the escalator barrel below Thameslink tracks and a larger one which today houses equipment for tunnel ventilation as well as emergency escape stairs.

In order to allow TBMs Phyllis and Ada to progress through the two deep shafts the lower portions were filled with concrete foam which the 150 metre long TBMs tunnelled through. The TBMs then travelled from the western ticket hall to the eastern ticket hall where their journey came to an end (see chapter 4).

114

At both Farringdon Crossrail sites the premium on space and existing underground services meant the diaphragm wall technique used at other stations had to be ruled out. Instead piling was used extensively to anchor and form the perimeter of the new structures.

At the eastern ticket hall the proximity of the sub-surface Underground line 10 metres below street level meant only part of the site could be excavated down to track level. Here one large shaft (trapezoidal in shape) was constructed to form the connection between ticket hall and platforms. Its size was dictated by the need to house tunnel ventilation, plant rooms, escalators and incline lifts, and as a means of emergency escape.

## Final fit-out

Next door to the 2011-opened Thameslink ticket hall at Farringdon, above the six escalators connecting to the Elizabeth line, more than 100 sections of diamond-shaped concrete panels were installed to create an angled roof featuring a ceiling design said to be inspired by the jewellery sold in nearby Hatton Garden.

This 'apse' ceiling may be enjoyed by passengers today but, if you listen carefully, you might hear the contractors who built it cursing. It consists of a 350 tonne, precast concrete structure which is suspended from a steel frame. To get it up required a 250-tonne steel support, likened to a gigantic jigsaw mat.[8] The temporary mat was removed once the permanent steel supports had been installed.

British artist Simon Periton was commissioned to deliver two art installations at Farringdon: in the western ticket hall large diamonds on glass panels appear to tumble around the escalators on the lower concourse level. A homage to the goldsmiths, jewellers and ironsmiths of nearby Hatton Garden, the two metre tall designs were digitally printed on to the glass wall panels.

---

[8] Farringdon 360 Virtual Tour and Exhibition: https://360.crossrail.co.uk/far/

At the other end of the station the glass wrapping three sides of the eastern ticket hall building is covered with a pattern that reflects the elaborate Victorian metalwork of the historic Smithfield Market opposite. After dark station lights illuminate the design from inside.

When the Elizabeth line was ready to open the partition wall in the integrated ticket hall off Cowcross Street was taken down. Passengers for Crossrail services now turn right and go down a short bank of escalators before walking round 180 degrees for the longer escalators to platform level.

## Totemic ambition

Much effort has been expended to minimise station clutter at Elizabeth line underground stations and this has resulted in the creation of Crossrail totems. These free-standing, four-way signposts give directions to the eastbound and westbound platforms, exits and connections with other lines. But they also uplight the ceiling, have loudspeakers for announcements, pull-out ribbons to allow areas to be cordoned off, emergency lighting and a handy electrical socket. Farringdon is a good place to spot them as it features 10 of the 75 installed across the central section Elizabeth line stations.

When Crossrail station designs were drawn up LED lighting was seen as a technology of the future. But during the development of the programme Elizabeth line lighting has become an obsession of designers working to minimise glare, avoid creating shadows and help CCTV cameras produce clear recordings. Naturally LED lighting has been embraced to cut the cost of energy bills at stations.

Even with low-energy lighting the heat generated is unwelcome underground. At Farringdon ceiling uplights are built into the totems but also feature in one and a half metre long uplight units built into the metalwork between escalators. Two and a half years worth of development has resulted in the design of a heat sink, similar to that you might find around a computer processor in a desktop PC, which draws heat away from underneath the LEDs in both lighting systems, limiting temperature around these to 35 degrees centigrade.

View of platform screen doors and destination indicator
at Liverpool Street station.

## LIVERPOOL STREET

| Role | Responsibility |
| --- | --- |
| Architect | Wilkinson Eyre |
| C138 design | Mott MacDonald |
| C210 advance works package 2 | Murphy |
| C212 demolition works | John F Hunt Demolition |
| C213 signalling relocation | Signalling Installation & Maintenance Services |
| C214 electrical and mechanical works | RGB Integrated Services |
| C215 cable identification | RGB Integrated Services |
| C216 utilities combined services package 2a | Laing O'Rourke |
| C225 site facilities | Select Plant Hire |
| C501 advance station works | BAM Nuttall/Kier |
| C502 main works | Laing O'Rourke |
| C503 EDF substation and advance works | Vinci Construction |

Liverpool Street links the Elizabeth line to the city of London, a bustling business district where corporate life is played out alongside trendy and fashionably understated Shoreditch and Spitalfields. Expected to be used by 124,000 passengers a day, this is another two-ticket hall station – Broadgate at one end is close to the Liverpool Street main line terminus and Underground station while, to the west, the new station joins up with (and takes over) the pre-existing Moorgate Underground station. These two ticket halls are connected by the Elizabeth line Liverpool Street platforms, the deepest on the new railway.

## Liverpool Street west

At Moorgate Crossrail works included a significant expansion of the existing ticket hall, connected to a new Moorgate shaft containing part of the main escalator bank from the ticket hall, emergency egress from the station platforms, fire fighting and passenger lifts, station and tunnel ventilation and part of the connection to the London Underground Northern line.

The Moorgate shaft is 42 metres deep. It is surrounded by London Underground's sub-surface lines to the north and Northern line to the east, plus about 50 buildings, some of them listed. With the shaft's east wall less than five metres from the northbound tunnel of the Northern line tunnel, it was essential to avoid ground movement during construction. The target was to limit horizontal wall movement to no more than 45mm but for the east wall this was set at a maximum of 30mm.[9]

While a careful-careful approach was clearly required the shaft needed to be fully excavated in time for the arrival of Crossrail's westbound tunnel boring machines Victoria and Elizabeth. However, extracting the foundations of a six storey 1970s building previously on the site took much longer than anticipated, delaying the start of diaphragm wall construction by 11 months. Without a new approach advance station works contractor BAM Nuttall/Kier would have been unable to meet the deadline.

---

[9] According to design consultant Mott MacDonald

118

The solution settled upon was a combination of high-tech modelling of possible ground movements with the markedly lower-tech 'observation-based' approach. This created enough confidence to proceed without needing to install temporary props – not only saving installation time but removing impediments to excavators working within the shaft.

Finding space for the shaft and to enlarge the Moorgate ticket hall was a challenge. After a fire on the roof of a demolition site in January 2011 enabling works contractor John F Hunt finished the demolition of 91-109 Moorgate at the end of May while demolition at 17-31 Moorfields, the cul de sac which runs parallel to Moorgate, was completed in the autumn. Even after this work existing structures restricted the ceiling height at Moorgate. Escalator shafts had to avoid a vast egg-shaped brick sewer tunnel designed by Joseph Bazalgette, the 19th century pioneer of municipal sanitation.

The premium on space informed the similar designs selected for both Liverpool Street ticket halls. These feature shallow, geometric folded ceilings, formed by ribbed pre-cast concrete panels, to break the perception of a flat ceiling and create a greater sense of space. The concrete mix includes the light-reflecting mineral mica.

By March 2016, at the Moorgate end of Crossrail's new Liverpool Street station, a new passageway had been built to make it easier for rail users to interchange between Elizabeth line and London Underground Northern line trains. Work on the Northern line link started in April 2014 and required particular skill and precision because of its proximity to underground utilities and structures at Moorgate. The six-metre diameter tunnel runs underneath the Circle, Hammersmith & City and Metropolitan lines and two escalators. It is also less than a metre below the Northern line's northbound tunnel.

During summer 2016 the final part of the new Moorgate ticket hall roof structure was completed. The roof sections form part of the new public interchange areas at Liverpool Street station and were installed following the demolition of the old Tube ticket hall, which was directly above the operational railway. This roof completion marked the end of the main heavy civil engineering works at Moorgate ticket hall.

Meanwhile, 30 metres underground, the two new Liverpool Street platforms were completed. As with its contract to build Custom House station, Laing O'Rourke made use of its factory near Sheffield to prefabricate more than 500 concrete platform components that were then transported to London, lowered down the station's main shaft and pieced together underground. The two new platforms took four months to install.

## Liverpool Street east

At the other end of the station there were plenty more underground structures to contend with including existing Tube lines and the old Post Office railway. But the most pressing issue was the mass of utilities that needed to move to allow construction to proceed.

Between the two ticket halls 22 telecommunications ducts, 18 high-voltage power cables and two water mains had to be diverted before the new eastern ticket hall could be built. Unfortunately they could not be relocated to nearby streets because of the shallow depth of London Underground tunnels. In response to this problem Mott MacDonald designed a 53 metre long cable corridor, a ventilated cut and cover tunnel five metres below ground, to accommodate the tangle of utilities. It became the first permanent sprayed concrete lining tunnel to be finished as part of Crossrail with a two-year diversion project involving careful negotiation and planning to ensure no interruption to water, power or telecoms. The cable tunnel now allows utility firms to carry out maintenance, or install additional services, without excavating roads and disrupting traffic.

Another major project at Liverpool Street out of view of passengers was construction of a 40 metre deep box below Blomfield Street. This houses ventilation and mechanical and electrical systems for the new Elizabeth line station as well as an emergency exit from Crossrail tunnels. This required the sinking of more than 250 piles up to 50 metres deep to create the Blomfield Shaft, Crossrail's deepest piled shaft.

Next to the box a new communications equipment room, EDF power substation and switch rooms for Liverpool Street London Underground were built and connected to the cable tunnel. This in turn allowed the demolition of the old substation to make space for the Broadgate ticket hall.

With the utilities dealt with surely construction of the new ticket hall could proceed? Only once archaeological digs by a team from the Museum of London to investigate a 16th century burial ground had been completed. Archaeologists uncovered 4,000 human skeletons between two and four metres below street level that had to be exhumed and relocated. Indeed, the new ticket hall would be built on the site of a burial ground used by the St Mary of Bethlehem psychiatric hospital (commonly known as Bedlam) from 1569 until the middle of the 19th century.

Following completion of another archaeological dig a temporary 42 metre deep shaft was dug at Finsbury Circus, roughly mid-way along the new station platforms, providing access to build the platform tunnels and cross-passages between the new ticket halls at Moorgate and Broadgate. This supported construction of 1.5km of sprayed concrete lining platform tunnels, concourse and passageways to link up the new ticket halls.

From the bottom of the shaft tunnels branch off in four directions. The platform tunnels were initially formed as six metre pilot tunnels which were then enlarged to the final diameter of 9.5 metres after TBMs Victoria and Elizabeth passed through the station. Once construction was complete the shaft was filled in and the gardens above reinstated.

By March 2017 construction of the Broadgate ticket hall, at the eastern end of the new Liverpool Street Elizabeth line platforms, was complete – crucially meeting one of the Crossrail project milestones relating to City of London payments towards the cost of the scheme[10] (the City reminds us of its contribution on the steps leading down from street level). Inside, the ticket hall ceiling is formed by 109 geometric light-reflecting concrete segments (supposedly referencing the pin-striped suits worn by financial bigwigs) designed to trick us into thinking it is not really a flat ceiling and the space below is more cavernous than it really is.

At the eastern ticket hall an incline lift alongside the escalators provides step-free access from street level to platforms. Natural light reaches this underground ticket hall through the five metre-high glazed 'canopy' entrance at Broadgate. At Moorgate the wide street level entrance helps daylight penetrate the ticket hall.

## High-level intervention

In addition to creating new underground facilities for the Elizabeth line at Liverpool Street work was also required at the main line, high-level station which is the start and finish point for a small number of peak time Shenfield services. Here platforms needed to be lengthened to accommodate full-length 9-car Elizabeth line trains.

Shenfield route Class 345s were initially delivered as 7-car sets which could be accommodated alongside existing platform faces. But, as these were extended to 9-car 205 metre long formations, modifications to the terminus were required.

Contractor BCM was appointed to extend platforms 16 and 17 with track removed and remodelled. The work involved modular platform extensions, slab track installation and changes to roof-mounted catenary. What was previously twin tracks between 17 and 18 is now a single line between a much-widened platform 17 and the now-fenced off former platform 18, retained to give access to staff facilities.

Most of the construction work for the project was completed at Christmas 2020 and Easter 2021 during two 10-day blockades. This met the 31 May deadline when 9-car trains – recently delivered to the Gidea Park sidings ready for use – were included in the new TfL Rail timetable.

---

[10] See chapter 2

Paddington station platform fit-out underway with
lily pad light fittings visible on the ceiling.

## PADDINGTON

| Role | Responsibility |
| --- | --- |
| Architect | Weston Williamson |
| Station and urban realm design | Gillespies |
| C130 design | URS Scott Wilson |
| C131 PIP design | Mott MacDonald |
| C251 Eastbourne Terrace utility diversions | Laing O'Rourke |
| C271 PIP enabling works | Carillion |
| C272 PIP main works | Carillion |
| C405 main works | Costain/Skanska |

Paddington Elizabeth line station is located next to the 19th Century
Grade I listed Brunel Paddington main line station with Crossrail

lines running almost parallel with the terminus platforms (although underground). In an attempt to build on Brunel's legacy the layout and structural elements of the new Paddington use the 10 foot imperial grid system to match Brunel's station.

Unlike other central London Crossrail stations this one has a single ticket hall, accessed via double storey escalators from each end of an excavated rectangular box approximately mid-way along the platforms. The sides feature pre-fabricated brickwork panels; eight flared elliptical columns clad in bronze support the roof and steel props above the escalators reinforce the box.

To excavate the new station required the site, previously housing the station's Departures Road taxi rank, to be cleared. But where would the taxis go? The answer – to the other side of the station as part of an ingenious plan to not only relocate taxis but build a ticket hall for the London Underground Hammersmith & City/Circle line platforms[11] and deliver a new canalside entrance for the main line station.

This scheme was known as the Paddington Integrated Project, drawn up to maximise use of a small portion of land between the main line station, the Regent's Canal towpath, and Bishop's Bridge Road, which runs over the canal to drop down by Departures Road. A new taxi ramp would connect Paddington's disused London Street Red Star parcels deck to Bishop's Bridge Road. The taxi rank and turnaround has been created on the parcels deck with lifts and escalators at the other end providing pedestrian access to and from the main line station. Underneath the ramp is the station's new 'Paddington basin' canalside entrance which also links to a new glazed London Underground station ticket hall above the LU platforms. Script-like motifs, emulating similar Brunel-era designs, connect the new and old buildings.

Given the need to clear the Crossrail box site the Paddington Integrated Project was sequenced to take place early in the

---

[11] These platforms,15 and 16, are in fact overground, beside the terminus platforms

programme, allowing taxis to be relocated to the new rank on 12 February 2012.[12]

Eastbourne Terrace, parallel to Departures Road, was then shut to allow the 260 metre long, 25 metre wide Crossrail station box to be built underneath; although the passenger entrance is on the station estate the station box extends underneath Eastbourne Terrace. The plan had been to complete the work in two halves to keep the road at least partially open at all times but ultimately it was decided that it would be more efficient to shut the entire road. Two of three lanes of traffic reopened in February 2014 with the Costain/Skanska contractual team further hemmed into what was already a narrow strip of a work site.

By 2013 excavations to create the station box were well underway. The roof slab was cast in September and at the west end an intermediate slab, a back-of-house area one level down, had also been completed. For most of the box this level was omitted to create the double height atrium for escalators down to the ticket hall. The concrete slab for this concourse level, which forms the roof of the Elizabeth line platform level, was finished by summer 2014. Precast square concrete panels featuring the 'lily pad' light fittings seen on the ceiling today were laid flat in the earth (on top of a protective covering) before the slab was cast and the platform level below excavated. Using these panels – 176 form the platform roof plus a further 96 at an intermediate level – allowed a high quality product to be created off site and quickly installed.

As the box was excavated contractors eventually exposed the top of the running tunnel segments. Lifting out the top most pieces was considered but rejected in favour of breaking through in order to retain as much of the tunnel structure as possible.

## Bakerloo connection

Construction of a deep level subway tunnel to provide a direct link between Elizabeth line and Bakerloo line platforms at Paddington

---

[12] Crossrail Paddington main works start in February, Transport Briefing, 30 January 2012

station began in spring 2014. Costing approximately £40 million the Bakerloo Line Link tunnel is 130 metres long, 25 metres below ground and has been designed to accommodate 5,000 people an hour interchanging between the two lines at peak periods. Escalators from the platform level link to a subway running underneath the main line station concourse to join the Bakerloo platforms beneath London Street.

Although the 2008 Crossrail Act specified construction of a direct link between the two lines, plans for the subway were amended to include a deeper level design that presented fewer risks to the Brunel-built main line station above. However, proposals to include a travelator in the tunnel were rejected as incompatible with the latest designs and the need to deliver the project in time for the scheduled opening of the Elizabeth line. After numerous schemes were considered by Crossrail design team Arcadis put forward an option which connected to the Bakerloo line platforms a third of the way along, rather than at one end, improving passenger flows and keeping costs down. The scheme was also developed to ensure the new tunnel did not impede London Underground plans for a future expansion of the Bakerloo line ticket hall.

The reworked tunnel design was not covered by provisions in the Crossrail Act, prompting London Underground to apply for a Transport & Works Act order to gain supplementary powers. Paddington was not the only place where Crossrail plans were refined after the Crossrail Act became law: TWA orders were applied for to provide access for Network Rail track work at Kensal Green and the train and maintenance depot at Plumstead sidings.

By March 2017 excavations had begun to build the foundations for the 120 metre-long glazed canopy, now situated along the side of the pre-existing Paddington station terminus, above the excavated box which forms the main entrance to the Elizabeth line. The canopy is made up of 180 glass panels, each weighing around a tonne, and featuring more than 25 different types of clouds as part of a design by American artist Spencer Finch. Although manufactured in Bavaria a mini mock-up was created in Barnsley to test installation of replica glass panels before the final designs were dispatched for Paddington.

With the panels installed and the Elizabeth line station complete what was Departures Road is now a landscaped pedestrian piazza.

Distinctive red and white props brace a newly constructed escalator box at Tottenham Court Road.

## TOTTENHAM COURT ROAD

| Role | Responsibility |
|---|---|
| Architect | Hawkins Brown |
| C134 design | Arup/Atkins |
| C208 demolition works | McGee Group |
| TCR east main works | BAM Nuttall/Vinci Construction |
| TCR east deep foundations | Bauer/Keller |
| C421 advance station works | Balfour Beatty/Morgan Sindall/Vinci Construction |
| C422 main works | Laing O'Rourke |

Tottenham Court Road station is pivotal to the business case for the Elizabeth line with the station soaking up demand for travel to and from London's most well known shopping district, a diverse range of businesses, tourist attractions and other leisure destinations. No surprise then that upon opening Tottenham Court Road was the most used Elizabeth line station.[13]

Like other central London Crossrail stations TCR has ticket halls at each end of the platforms – platforms which run parallel to Oxford Street above. The western ticket hall is accessed via Dean Street; this follows a decision not to have the facade on Oxford Street in order to avoid pedestrian congestion around the entrance (a familiar sight at Oxford Circus further along the road). Not only is the western ticket hall well sited for shoppers but it also provides access to the Soho district where many media and production companies are based.

The eastern ticket hall is located at a crossroads at the east end of Oxford Street with Tottenham Court Road heading north towards Euston and Charing Cross Road south towards Charing Cross rail terminus and Leicester and Trafalgar Squares. (A diagonal 'Shibuya' zebra crossing was proposed for this location but has yet to materialise). This is the site of the pre-existing Tottenham Court Road London Underground station serving the Northern and Central lines. Centre Point, a 34 storey tower completed in 1966, is a useful landmark and its foundations have necessitated the construction of the only curved platform on the Elizabeth line.

Here delivery of the new Elizabeth line station differed from other sites. While the Dean Street development is an all-new Crossrail facility, the eastern access point was delivered through a £500 million overhaul of the existing station. Although it built in capacity for the Elizabeth line this was a London Underground project programmed to go ahead with or without the new railway.

Because the Tottenham Court Road Tube station development got underway in advance of Crossrail it provided an opportunity to glimpse

---

[13] Transport for London press release, 1 February 2023

the scale of construction to come and attempt to understand the interdependent components of this densely populated part of central London. Following the completion of the King's Cross Underground station upgrade in autumn 2010, the expansion of TCR Underground became Transport for London's most advanced capital project.

## Eastern ticket hall

The LU station main redevelopment began on site in 2010 (relocation of utilities and other enabling works had been ongoing for more than three years) with a 2016 scheduled completion date. A joint venture of BAM Nuttall and Taylor Woodrow, which became part of Vinci Construction, was appointed main contractor for the scheme in 2009. It was agreed that the joint venture would take on all civil engineering required for Crossrail on the TCR east site.

Problems presented by the existing station included columns in the way of ticket gates, narrow staircases leading to narrow pavements; fundamental issues at every level which meant keeping the station functioning smoothly every day a challenge. The arrival of Crossrail was expected to see passenger numbers increase by a quarter from 150,000 a day to 200,000 and this led to plans to expand the ticket hall below Charing Cross Road to be at least five times the size of the one it would replace.

Initial work involved the controversial demolition of the Astoria (a dilapidated but iconic music venue), removal of the water feature and sculptures[14] at the foot of Centre Point, and the closure for four years of the top of Charing Cross Road where it joins Tottenham Court Road and meets the east end of Oxford Street. This allowed sites on either side of the road to be joined up to create a vast construction compound. Fans of Robert Galbraith's 'Strike' books will be familiar with the interminable roadworks around the lead character's Charing Cross Road haunts that mirror the Tube station redevelopment.

---

[14] Some of these have been relocated to the middle of a wood at Hooke Park in Dorset. I visited them in 2023

Construction of the eastern ticket hall underway
at the base of Centre Point (top right).

Piling started in 2010 to create the underground walls which
would enable the ticket hall and pedestrian links to be excavated.
A northern line escalator box was constructed to connect three
new escalators from the southern side of the new ticket hall to the
Northern line platforms, relieving pressure on the existing escalators.
Reaching 22 metres below ground, contractors then tunnelled from
the bottom of the escalator shaft to create a lower concourse and the
connecting passageways to Northern line platforms.

In preparation for the new Northern line link work began at platform
level in April 2011 with trains not stopping at the Underground
station for eight months. This allowed work to realign the tunnel
rings to provide sufficient space between for the lifts and stairs to the
new lower concourse above. Rather than go to the trouble of boring
new tunnels, the profile was altered to make them narrower – ellipse
shaped rather than circular. This procedure had previously been
successfully used to install Tottenham Court Road escalator number

six but, given that it required sections of platform to be cut away to gain access to the tunnel rings, it was far from straightforward.

A new addition at the Tottenham Court Road east site was the Goslett Yard box, on the Soho side of Charing Cross Road, where the existing station would connect to the Elizabeth line. The box, which now houses more escalators, was constructed top down with 44 concrete diaphragm wall panels cast and 11 plunge piles going down 50 metres – without hitting anything. The westbound Elizabeth line tunnel goes through the bottom of the Goslett Yard box, just under the new Northern line escalators.

On the eastern site at Goslett Yard, archaeological investigations unearthed a vaulted chamber. This was believed to be a cistern which had been filled with more than two tonnes of ceramic and glass vessels when the structure went out of use. The ceramics are associated with pickle brand Crosse and Blackwell, back then a flourishing Victorian company which was known to have occupied the site until the 1920s.

One further excavation at Tottenham Court Road that few passengers will use is the Falconberg Shaft, an emergency escape route from station platforms via a 20 metres deep shaft.

The new eastern ticket hall, and the new escalators to the Northern line, opened in early 2015. This marked the completion of a significant portion of the station upgrade and signalled the start of the next phase – which included demolishing the existing ticket hall to allow the new one to be enlarged and completing the refurbishment of the old parts of the station.

The next phase would also see Central line platform walls reprofiled – a process involving the removal of cast-iron curved sections of the platform walls which were replaced with vertical steel sections. The aim was to create sufficient space between the platforms for the construction and breakthrough of new staircases as well as a new lift shaft. These connect the reprofiled platforms with the new ticket hall via a newly mined interchange passage. During most of 2015 Central line trains did not stop at the station so as to let this work proceed.

## Signature shapes

Tottenham Court Road Tube station has multiple entrances. While there is no longer a fountain at the base of Centre Point there is more space for pedestrians with new prismatic glass entrances, which opened to passengers in 2015, leading on to a piazza area. Downstairs the new ticket hall is clad with glass panels featuring Daniel Buren's 'Diamonds and Circles' artwork as well as a three-dimensional sculpture. London Underground was criticised for destroying some of the Eduardo Paolozzi mosaics installed at the station in 1984.[15] However, more than 95% of the mosaics on the Northern line platforms have been retained and restored. Part of the former Paolozzi escalator arches was donated to the Edinburgh College of Art.

In February 2017 Transport for London announced the completion of the London Underground-led transformation of Tottenham Court Road eastern station. Elizabeth-line ready it became the 71st Tube station to offer step-free access with three new entrances (and one further refurbished entrance), a massively expanded ticket hall, eight new escalators and six new lifts.

## Western ticket hall

For the western ticket hall there was no existing station or railway to work around but similarly dense urban development necessitated careful planning.

The first step was to clear buildings from the area bounded by Dean Street, Diadem Court, Great Chapel Street and Oxford Street. By 2011 relocation of utilities was largely complete allowing advance station works – piling and construction of the diaphragm walls – to begin. The station box would then be excavated top-down with concrete slabs cast for each level.

Excavations had previously revealed fragments of Roman pottery and the remains of an earlier thoroughfare called Parkers

---

[15] Paolozzi arches at Tottenham Court Road 'already demolished', Architect's Journal, 30 January 2015

Lane. Archaeologists also uncovered the remains of old cellars which suggested the ground level in Great Chapel Street was approximately three metres higher than it had been in the 17th century.

Below ground the western ticket hall has five basements – B1 to B5. At ground level there are two buildings, reflecting the street pattern where Fareham Street bisected the site. To the north of Fareham Street – which has been realigned – is the Dean Street building forming the entrance to the western ticket hall and to the south the Fareham Street building, housing most of the operational facilities – including ventilation shaft, emergency access and plant rooms.

The decision to switch the station entrance to Dean Street made space for a ground floor retail frontage directly on to Oxford Street and supported plans to part-pedestrianise Dean Street – allowing passengers emerging from the station to reorientate themselves and also use the entrance as a meeting place.

By the end of 2016 the western ticket hall concrete super-structure was complete. Transport for London agreed to pay main contractor Laing O'Rourke more than budgeted to ensure work on the oversite development was completed by the time the central section of the Elizabeth line opened. Not doing so would have risked possible future disruption to passengers as well as leaving a 'gap-tooth' street-scape with the exposure of station ventilation structures in a prime site next to Oxford Street.

Underground at Tottenham Court Road 900 metres of floating track slab was installed – one of two sections of the Elizabeth line that required floating track to minimise noise and vibration from the operational railway because live music venues and recording studios are located directly above.

By November 2017 installation of the glazed walls wrapping the Dean Street western entrance to the new Tottenham Court Road station was complete. The five metre-high panels allow natural light to filter into the ticket halls during the day and station lighting to illuminate the street at night.

## Final effect

Of all the dual entrance stations it is at Tottenham Court Road where the finished ticket halls provide the most striking contrast. White glass and stainless steel dominate the colour palette at the eastern entrance to complement the 1960s design of Centre Point tower. Glazed elevations to the ticket halls together with clever lighting reinforce the presence of the station at night and provide a connection to natural light during the day.

The western ticket hall in Soho is described as "dark and cinematic"[16], reflecting the nocturnal economies that characterise the area. At this site black is the colour of choice for the glass and stainless steel inside the station. Below ground, distinctive 'drum' light fittings include acoustic absorbers to keep down noise and echoes. Intended to reference theatre-style lighting, the drums are set between deep roof beams.

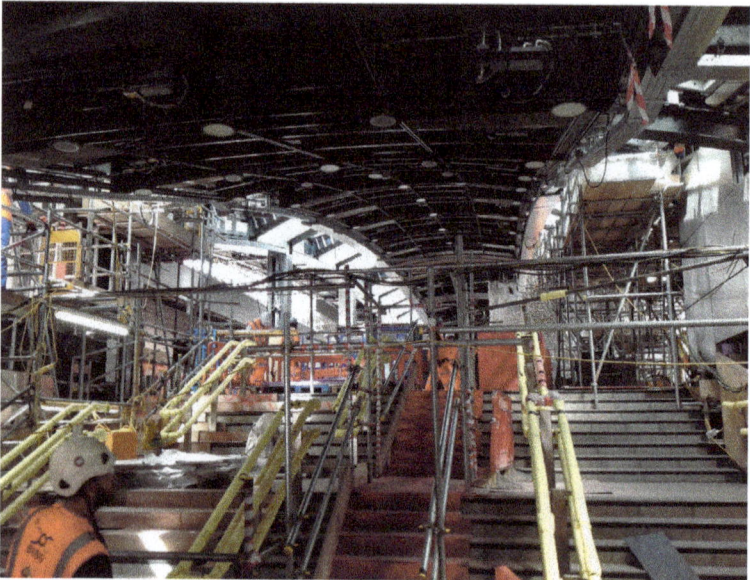

At Whitechapel a new concourse was built above the existing London Underground and Overground railways.

---

[16] Hawkins Brown, https://www.hawkinsbrown.com/projects/tottenham-court-road-elizabeth-line/

# WHITECHAPEL

| Role | Responsibility |
|---|---|
| Architect | BDP |
| C140 design | Hyder Consulting |
| C217 civils WP2 | Carillion |
| C244 East London line steel working deck | Kier |
| C245 civils, demolition and utilities | Murphy |
| C511 advance station works | BAM Nuttall/Kier |
| C512 main works | Balfour Beatty, Morgan Sindall, Vinci |

It wasn't supposed to look the way it does today. Originally Whitechapel Crossrail station was designed to have its main entrance on Fulbourne Street, leading to a lengthy concourse above the surface level Hammersmith & City/District line Underground tracks which would eventually connect with escalators down to the Elizabeth line.

But in June 2010 new designs were made public, revealing that this station would look like no other on the Elizabeth line. Architect BDP chose to switch the axis of the concourse from east-west to north-south. Instead of sitting above the H&C/District it would be built above the East London line, now part of London Overground, which sits in a cutting at roughly ninety degrees to the other pre-existing, surface-running railway.[17]

BDP had a persuasive list of reasons for the rethink. Passengers would not have to walk so far from the station entrance to reach Crossrail platforms (and vice versa). Fewer turns would be needed and more of the journey through the station would be within daylight. A promise to reduce station costs by £50 million probably helped secure sign off for the amended plans.

---

[17] At Whitechapel the Underground lines cross over the Overground

Consequently the Elizabeth line station retains the Underground station's Whitechapel Road entrance, which dates from 1876. The old ticket hall inside has been demolished; passengers today climb a short flight of stairs (or take one of the station's 10 lifts) to an upper concourse above the District/H&C (from here stairs provide access to platform level) before descending to a lower concourse, perched on steel struts resting on the brick arches of the London Overground cutting. Timber-effect ceiling ribs hide the pipes and wires installed above. The lower concourse extends under Durward Street after which a 130 degree turn provides access to a long bank of three escalators, 30 metres down to the Elizabeth line.[18] Beyond the escalators a second entrance/exit has been created at the north end of the station.

So where for other new Elizabeth line stations we have talked in terms of ticket halls, Whitechapel is about an elaborate railway overbridge, (the design allows daylight to reach the London Overground platforms below), combined with shafts linking to platforms.

Durward Street shaft is the main deep level structure, sliced diagonally by the escalators that link Crossrail platforms with overbridge. To the east of the platforms is the Cambridge Heath shaft, built to provide emergency access. I have been told this could be upgraded to create an additional passenger entrance/exit but there are no plans to do so.

As well as providing access to the three sets of railway lines served by the station the overbridge doubles up as a new public route between Whitechapel Road and Durward Street, overcoming the local barrier created by the H&C/District line. A glass screen separates ticket holders from pedestrians not using the station.

## Building the station

To prepare for construction of the new station a working platform was built over the East London line. This allowed work to remove

---

[18] In perhaps the most tenuous Crossrail design explanation I have heard the concrete lining to the escalator hall is said to be based on a notation of bell resonance referencing the bell foundries that used to exist in the area

an old footbridge and then construct the new overbridge and Durward Street shaft without the risk of tools or debris falling on to a passing Overground train.

Excavation of platform tunnels began from a temporary Brady Sheet shaft to the west end of the Cambridge Heath shaft in June 2011. By this point construction of diaphragm walls at both the Cambridge Heath and Durward Heath shafts was underway.

To make space for the shafts and site compounds extensive preparations were required including building a new access road for the Albion Brewery and filling in a well beneath the brewery basement. At Durward Street a storm water sewer had to be strengthened and Crossrail was required to construct a new building for Swanlea School plus undertake modifications to the nearby sports centre.

By March 2013 the Cambridge Heath shaft, located towards the eastern end of the future Elizabeth line platforms, was 22 metres deep. The 30 metre diameter shaft, providing ventilation and emergency access would eventually go 35 metres down.

In July 2013 fit out of the Durward Street shaft, a 30 by 60 metre box north west of the new station concourse/overbridge began. Main station contractor BBMV took over from the BAM Nuttall/ Kier station advance works team which left the Durward Street site following the completion of the capping beams which joined the 53 diaphragm wall panels of the Durward Street shaft together. BBMV was then responsible for excavating 30 metres down and installing three levels of support structures to reinforce the shaft walls. A concrete base slab was cast at the bottom of the shaft.

Platform tunnels at Whitechapel were constructed ahead of the arrival of TBMs Victoria and Elizabeth in summer 2014. By then, at the eastern end of the new platforms, work was well advanced on the 35 metre deep Cambridge Heath emergency access and ventilation shaft. Passageways were mined to connect the shaft with a cross passage linking the southern and northern platform tunnels.

To fit out and connect the new concourse the Underground/
Overground entrance on Whitechapel Road had to close for around
three years. A temporary station entrance was established on
Durward Street, the other side of the sub-surface District and
Hammersmith & City lines, accessible from Whitechapel Road via
Court Street and Fulbourne Street. In preparation for this Court
Street received a makeover with new paving and lighting.

When the Whitechapel Road entrance closed in January 2016
and the temporary one opened some station users praised the
spaciousness of what they thought was the new entrance. But when
the renovated Whitechapel Road station entrance reopened it was
much more dramatic - old ticket office and back of house partitions
had been removed to make the most of the entrance through the
historic facade and which now leads into a substantial lobby area
before the station gateline.

By February 2015 Crossrail's tunnel boring machines had passed
through Whitechapel. Given the severe constraints of the
Whitechapel Durward Shaft site (flats, sports centre, school), which
allowed limited space for heavy machinery, a decision was taken to
excavate the escalator shaft bottom-up, rather than employing the
standard top-down approach. This required the use of an uphill
excavator, more commonly found in coal mines, and was the first
time this uphill mining technique had been used on a UK rail project.

At the end of 2016 construction of the new concourse overbridge
was well underway with the structure, suspended above the East
London line, clearly recognisable. The Cambridge Heath Shaft was
structurally complete and ready for fit-out.

## Paying the price

The Whitechapel station design is very clever but as the station build
progressed was there a sense that perhaps it was too clever? The
constrained work sites and access difficulties made Whitechapel one
of Crossrail's most challenging sites and among the last to finish.
Visiting the site it seemed more of an overground scheme with less of
the extensive underground activity on view than at the other big

station construction sites. But of course Whitechapel's platforms are on the same scale as the other stations and so required the same construction mobilisation, even if worksites above ground were not conducive to this. It also has a crossover for trains with these crossover chambers at Whitechapel among the biggest underground structures anywhere on the Crossrail project.

Whitechapel certainly had its fair share of problems. The 2019 report by the National Audit Office examining Crossrail contracts found the cost of C512, the main works contract, increased by nearly 500% as a result of difficulties building around existing London Underground and overground lines and station architecture.[19] When I visited the work site in summer 2019 fit-out was nearly complete; a few missing light bulbs above platform screen doors, ceiling ribs and escalator panels but otherwise Whitechapel appeared close to being finished.

Walking outside the station I was shown the green sedum roof (to improve drainage and reduce noise) above the lower concourse and the Cambridge Heath shaft – now clad with striking vertical bronze, gold and black strips. These can be admired from the beer garden of The Blind Beggar, the pub where 1960s gangster Ronnie Kray shot and killed a rival.

---

[19] See chapter 11

Woolwich was a late addition to Crossrail plans and part funded by housing development around the station.

## WOOLWICH

| Role | Responsibility |
| --- | --- |
| Architect | Weston Williamson |
| Contractors for Berkeley Homes | Balfour Beatty Ground Engineering Services, Byrne Bros, O'Keefe Construction, Gallagher |
| C530 main works (box fit-out) | Balfour Beatty |

Woolwich Elizabeth line station is located at the historic site of the Royal Arsenal in south east London – a dockyard established by King Henry VIII which became increasingly important as an ordnance factory in the 17th century, housing a military academy and Royal Laboratory as well as a new brass gun foundry in the early 18th century. The Royal Arsenal expanded rapidly in the mid to late 19th century, reaching peak activity during the First World War.

Entry to the station, which has a single ticket hall, is via a 30 metre wide bronze clad opening on to Dial Arch Square, a green space flanked with Grade I and II listed buildings. Linking the station to the area's history, two metal murals provide a contemporary representation of the bronze memorial plaques that were minted there during the First World War. This cladding incorporates images of Britannia and the Lion which featured on what came to be known as the 'Dead Man's Penny', the ceremonial coins given to the families of soldiers who gave their lives in the Great War. More than one million of these plaques were cast at the Royal Arsenal.

The Elizabeth line's only all-new station south of the River Thames certainly shores up the regeneration credentials of the Crossrail programme but it nearly didn't happen. When the Crossrail bill was laid in parliament in 2005 it envisaged trains travelling from Custom House to Abbey Wood without stopping.

As recounted in chapter 2, during the passage of the Crossrail bill the government and the organisation which became Crossrail Ltd argued against the station proposal amid fears that Crossrail costs were already rising. But determined lobbying from the Royal Borough of Greenwich and Greenwich and Woolwich MP Nick Raynsford won over doubters; in March 2007 Transport Secretary Douglas Alexander announced a funding deal that would allow the station box to be built.

Adding the Woolwich stop to Crossrail was estimated to add £186 million to the price of delivering the programme. Berkeley Homes funded construction of the box - a structure 18 metres deep, 26 metres wide and 256 metres long – to support its extensive building programme in the vicinity. The deal allowed the government to claim that the cost of Crossrail to the taxpayer had not risen, despite the inclusion of an extra station.

This agreement got the structure for the station built but a stand-off over who should pay for fit-out meant it wasn't until 2013 that Berkeley, Transport for London and Greenwich council agreed to share the cost of fitting out the station box. Commitments were only made after Crossrail Ltd warned that it needed an instruction to proceed with fit-out if works were to be delivered in a cost efficient manner and in time for the start of Crossrail train services - then expected in 2018.

## Strong arm tactics

The station box actually resembles a dumb bell in shape – at the western end a rectangle marks out the perimeter of the ticket hall while a smaller rectangle at the far end houses emergency escape stairs from platform level. In theory this could be upgraded to become a second ticket hall, similar to the central London Crossrail stations, should passenger demand ever warrant the additional investment. Linking the two is another rectangle, this time long and thin to reflect the footprint of Crossrail lines and platforms.

With time tight to construct the station box, in advance of the arrival of tunnel boring machines Sophia and Mary in 2013, Berkeley decided to speed up construction by splitting the dumb bell box in two by creating a temporary concrete wall, initially underground, that allowed the western half – the ticket hall end – to be dewatered and dug out while diaphragm wall building for the eastern half continued.

Another Berkeley decision was to excavate three metres of ground across the entire site before starting construction of diaphragm walls. The logic was that, while this would create extra work, it would avoid interruptions later on if, as subsequently materialised, evidence of previous human activity was discovered on site.

Removal of this surface layer allowed clearance of all signs of human habitation, the diversion of utilities and preservation of items of archaeological importance. With the knowledge that the ground remaining had not been touched by humans, contractors could use heavy construction machinery without having to worry unduly about what it might encounter.

Another precautionary measure was the creation of a contiguous piled wall around the site to shore up the nearby A206 while temporary sheet piles were installed to support the foundations of listed buildings on the north side, including the old Royal Carriage Factory.

Further work required before main construction could start included rerouting a 1.2 metre diameter sewer. Shafts up to seven metres deep were built down into which a mini tunnel boring machine was lowered. Controlled remotely, it was used to divert a 400 metre long section of sewer around the station box site. New pipes were pushed into place behind the machine.

Following the enabling works Balfour Beatty Ground Engineering Services began building the diaphragm walls to form the station box. Rigs fitted with hydraulic grabs chewed a slice out of the ground with the trench created filled with bentonite to support the sides. Steel reinforcement cages were lowered into the void into which concrete was poured, displacing the bentonite. Once set a 25 metre deep wall remained.

The station box construction at Canary Wharf revealed the Thanet sands which are usually hidden under London but at Woolwich a less construction-friendly layer was encountered. Bullhead beds comprise a thin layer of dense brown and black sandy gravel and flint cobbles and are problematic because they are porous. Dig a hole in the ground and fill it with bentonite and rather than the liquid clay staying put until the concrete has been poured in, it leaks out into the surrounding ground. Without the bentonite the hole will collapse so it's vital to ensure that the liquid can be retained for as long as needed. Fissures within the chalk seam underlying the bullhead beds were also a concern for bentonite loss.

To solve this issue grout was injected into the ground along the perimeter of the station box. This filled any cracks and fissures in the Bullhead and chalk layers to prevent bentonite seeping into the surrounding earth.

With diaphragm walls constructed, eventually forming the dumb bell shape, work to excavate the western half of the box began. A concrete slab would later be cast 18 metres below ground but, given the proximity of the station to the Thames, the diggers encountered the water table before reaching this depth. This required a complex arrangement of pumps to keep the work site drained and temporary steel props to brace the box.

**On site at Woolwich**

When I visited the Woolwich station site in 2012 Berkeley Homes project manager Shane Taylor summarised the construction sequence as follows:

1. Site clearance
2. Excavation of top three metres of site
3. Initial site excavation completed
4. Diaphragm wall construction with props installed as first half of box is excavated
5. Props installed in central section of box. Piling for oversite development
6. Props installed at eastern end and second layer of props installed at western end
7. Concrete slab cast at track level – bottom row of props removed
8. Concrete slab (roof) cast over platforms
9. Box construction complete. Remaining props removed once oversite development in place

## Going overground

Crossrail's nine new central section stations are a revelation and show just what can be achieved, albeit at a price. But the Elizabeth line is more than an underground railway with a further 32 stations – overground except at Heathrow – completing the route at opening. Here it would fall to Network Rail to take the assets it had inherited and deliver an extensive programme of upgrades specified by the Crossrail Act.

# 6. THE OVERGROUND

Alongside the pizazz promised by new Crossrail tunnels and stations in central London, surface Elizabeth line works are sometimes overlooked. But the scale of this activity overground was significant: Network Rail presided over the upgrade of 27 stations, 61km of track, 179 points and crossing units plus around a thousand overhead line gantries. That's alongside building two major new structures and upgrading existing signalling. The work also included electrification of the route from Heathrow's Airport Junction to Reading and interfaced with many of the other big rail enhancement programmes of the time – remodelling Reading station, Intercity Express and Thameslink.

In broad terms Network Rail was responsible for the design, development and delivery of the parts of the Elizabeth line which are above ground. Unlike the tunnelled sections of the route, most of these works needed to take place on a live railway.

But the organisation's remit went further. In 2011 Network Rail principal programme sponsor Rupert Walker explained to me that the organisation had four Crossrail roles. While it was Crossrail Ltd's delivery partner for the overground, 'on-network' works, it was also infrastructure manager of the national network and Crossrail could not be allowed to detract from its day to day business of enabling trains to run and passengers to get to and from work.

Network Rail was also a landlord. In Crossrail terms that meant deploying teams at Westbourne Park, Pudding Mill Lane and Plumstead to relocate signalling equipment and provide access for tunnel boring machines. And as owner of sites such as Paddington and Liverpool Street stations it had a duty to check the development of Crossrail plans to ensure that no big holes were likely to open up when TBMs start digging under a main line terminus. Finally, it had a role as an expert; if Crossrail wanted advice on a rail issue then Network Rail was the logical go-to source of information.

Overground the Elizabeth line network consists of three sections which are joined together by the newly built twin tunnels through central London and Docklands. To the west the on-network route runs from the Royal Oak Portal west of Paddington to Reading with a spur connecting Heathrow Airport. To the east there are two branches – one emerging at the Plumstead Portal and continuing south east beside the North Kent Line to Abbey Wood and one emerging at the Pudding Mill Lane Portal and which uses track along the Great Eastern Main Line to head north east towards Shenfield. All three of these overground sections existed before Crossrail but have since been modified to accommodate the Elizabeth line service.

When construction of Crossrail got underway early activity naturally centred on the new sections of tunnelled route where all-new infrastructure needed to be built. With Crossrail's overground sections already carrying trains – albeit not in an Elizabeth line configuration – there was at least initially a sense that this part of the programme could wait. This was a shame as in many cases work, when it did happen, took longer than planned, and Network Rail struggled to meet delivery deadlines.

Nevertheless, despite a slow start, by the end of 2011 detailed design work for nearly all Network Rail Crossrail schemes had been completed with enabling works for the big overground projects getting underway.

## Western (Royal Oak Portal to Maidenhead)

Of the three Elizabeth line overground sections the western route was where the bulk of Network Rail's Crossrail upgrade took place.

Major infrastructure projects at Stockley and Acton, electrification, a comprehensive package of station enhancements – plus wide-ranging track and signalling changes – occupied engineers and specialist teams for years. Twenty bridges needed to be demolished and rebuilt to provide clearance for the new overhead electric lines required by the Class 345 Elizabeth line rolling stock. A multi-phase programme of signalling upgrades was needed to provide computer based interlockings all the way from Paddington to Reading. Completed in 2015, this allowed signalling along what would become the Elizabeth line western route to be transferred to the Thames Valley Signal Centre at Didcot.

The western surface section of the Crossrail route from the Royal Oak Portal through to Maidenhead was where on-network activity was concentrated and had to be planned to integrate with other major upgrades on the Great Western route out of Paddington including electrification, emerging plans to fit the European Train Control System (ETCS) and the Intercity Express Programme.

Early on Vinci Construction was confirmed as winner of both the inner and outer western station packages with a combined value of around £100 million and covering main works at at least 13 stations. New station buildings were required at Acton Main Line, Southall, West Ealing and Hayes & Harlington. A major renovation of Ealing Broadway would deliver an enlarged ticket hall, new footbridge, new lifts and pedestrianised forecourt.

Lifts and platform extensions were planned for West Drayton station while Maidenhead station would get new lifts, a new platform and platform extensions. Today Maidenhead provides sidings for Elizabeth line trains and staff accommodation for drivers and cleaners. Platform extensions were also needed at Langley, Burnham, Slough and Iver.

With main Crossrail on-network works getting underway in 2012 the first Crossrail blockade was for Slough resignalling. A Christmas possession allowed the functions of one of two Slough signal boxes to be transferred to the Thames Valley Signalling Centre. In May Network Rail awarded Signalling Solutions a £5 million contract to replace 13 existing solid state interlockings with four of Alstom's Smartlock 400T interlockings. Acton and Stockley possessions followed with the scale of these projects meaning it would take years rather than weekends to have the new infrastructure up and running.

As part of a shutdown to support the (non-Crossrail related) Reading station redevelopment a Christmas closure also provided the opportunity for Network Rail to demolish Dog Kennel Bridge just west of Iver, one of Brunel's less remarkable structures and which had insufficient clearance for Crossrail electrification. British Rail had previously secured legal powers to demolish the bridge in 1992 as part of a Great Western enhancement scheme that was never implemented.

At Old Oak Common the track layout for connecting the new Crossrail train depot to the Great Western corridor was revisited in the light of the evolving Intercity Express Programme, which now makes use of the former Eurostar North Pole depot on the other side of the railway. Arup, which was appointed to work on the track design for both projects, was charged with ensuring there were no conflicts between Crossrail's designs and IEP requirements.

Other changes included the Hanwell permanent way scheme, which was developed to reduce the risk to operational capacity posed by freight trains travelling from the Down Relief Line to Southall Down Yard. At West Ealing a long-running proposal to replace Greenford-Paddington trains with a Greenford-West Ealing shuttle was made possible by the construction of a bay platform and the installation of lifts to ensure shuttle users have step free access to and from the frequent Elizabeth line central London connections. Getting rid of the lightly used 2-car Greenford-Paddington trains freed up valuable paths for longer trains on busier routes.

By early 2014 Crossrail Ltd had submitted proposals to the London Borough of Ealing to upgrade Ealing Broadway station. The plans would see the entrance and ticket hall replaced with a more spacious, glass-fronted structure. New lifts and staircases would make it easier for passengers to get around the station and the entrance area would be twice the size of the old one, allowing a wider gateline and the dated facade to be replaced. Staircases to platforms 1 and 2-3 were redesigned, platforms extended and waiting facilities much improved.

More radical – and expensive – plans for Ealing Broadway station had previously been proposed with suggestions that a large oversite development could have helped finance the installation of escalators and other improvements. But following much discussion with council planners agreement was reached on a less ambitious scheme that increased station capacity and provided step free access while still being deemed affordable.

Several of the west London stations would gain new ticket halls in preparation for the opening of the Elizabeth line. Glass and steel structures designed by Bennetts Associates have become a hallmark of this part of the route with the architect's designs applied to stops including Southall, West Ealing and Hayes & Harlington.

By early 2015 planning applications had been submitted for many of the western stations. The Acton Main Line overhaul included new lifts to provide step free access to all platforms, platform extensions plus new canopies, lighting, customer information screens, station signage, help points and CCTV. At Hayes & Harlington the new Bennetts-designed station building replaces a commercial property and links to a new footbridge with four lifts providing step free access to all platforms. These have been extended and the canopies replaced.

Shortly afterwards proposals for a similar overhaul of West Drayton station were put forward. A new glass and steel extension now provides an additional entrance to the station as well as a covered walkway between the existing building and a new footbridge. The designs sought to ensure that the Victorian building remains the focal point of the station while providing modern facilities including three new lifts to platforms.

Despite having awarded Vinci a design and build deal for the western stations work in 2013 Network Rail subsequently chose to let new contracts for the construction package. While Crossrail teams in central London were fully mobilised it was not until May 2019 until the green light was given to deliver the long-planned western station enhancements. Network Rail split the work into two packages: Hochtief took on the West Drayton, Hayes & Harlington and Southall upgrades while Graham delivered improvements at

Acton Main Line, West Ealing and Ealing Broadway. Network Rail said it had started work on foundations and steel frames for new footbridges and lift shafts the previous year.

## Acton dive-under

In addition to track, signalling and station work two distinct infrastructure enhancements were required on Crossrail's western route.

Construction of the Acton dive-under and associated track improvements allows trains bound for Paddington to go under freight trains from Acton Yard, avoiding delays, increasing freight capacity and improving reliability. Track was designed to ensure extra-long freight trains from the yard can gather speed before joining the main line so that westbound trains behind are not delayed. Prior to its construction passenger trains passing through Acton were regularly held up by long freight trains entering and leaving the yard. Without the dive-under freight trains to the yard would use up at least one Elizabeth line train path an hour.

The completion of enabling works at each end of the yard to isolate it from the operational rail network meant work on the project could continue during the 2012 Olympic and Paralympic Games, even when a moratorium on railway enhancements was in place. Network Rail subsequently appointed BAM Nuttall as main contractor with a package of work valued at £22 million.

Main construction of the dive-under began in October 2013. More than 34,000 tonnes of earth were excavated from a five metre wide site, surrounded by the operational railway, with freight trains passing on one side and passenger trains on the other. Over 1,400 piled foundations were installed and construction involved nearly 40,000 tonnes of concrete and the installation of 730 metres of track.

BAM organised the project into four stages with the first two – building the first half of the dive-under and flood prevention works – taking place simultaneously. These were followed by construction of the second half of the dive-under and the remainder of works, including laying slab track. Bauer was piling sub-contractor for the project.

Following completion of the main structure in July 2016 electric overhead wires to power trains and signalling were installed before the first test trains used the dive-under towards the end of the year. The dive-under came into full operation at the start of 2017.

## Stockley flyover

The other major project Network Rail was charged with delivering on the western route was the Stockley flyover enhancement where the Heathrow spur branches off the Great Western corridor. This now provides a connection between the airport branch on the south side and the Crossrail lines on the north side, routing Elizabeth line services up and over the Great Western main lines.

At Airport Junction Network Rail's brief was to double the previous partial grade separation and increase the existing capacity for six trains an hour in each direction (four Heathrow Express, two Heathrow Connect) to eight (four Heathrow Express, four Elizabeth line).

Enhancements to the Stockley structure were required to carry trains from Crossrail lines on the northern side of the GWML corridor to the Heathrow spur on the southern side. The structure, which includes a new single track viaduct for all trains from Heathrow towards London, also allows Elizabeth line and London bound Heathrow Express trains to access the airport without delaying trains on the main lines.

Network Rail awarded Carillion the £36 million main civil engineering contract to expand the flyover. Initial work involved construction of a new embankment on the Heathrow side and concrete pouring for the central pier, built between existing lines, ahead of the launch of the new flyover deck.

On 21 May 2014 Network Rail and main contractor Carillion completed the installation of a 120 metre long Stockley western flyover across the railway. The launch of the thousand tonne bridge began on 13 April and involved incrementally jacking the curved steel truss structure across the four rail tracks to/from the west. This was the largest single span rail crossing to have been installed anywhere to the west of London since the days of Isambard Kingdom Brunel.

To start with the bridge was assembled on a new west abutment which had been constructed alongside the embankment where the Heathrow spur branches off the main line before entering tunnels to the airport. Two hundred and seventy five pieces of steel, supplied by Martifer, were brought to site. Once bolted together five hundred tonnes of concrete were used to cast the first part of the bridge deck. This then served as a counter weight as the flyover structure was gradually jacked across the Great Western.

From the west abutment the bridge had to be manoeuvred across the two fast lines to a centre pier and from there across the two reliefs to an east pier on the other side of the rail corridor. PTFE slide plates on top of these two piers provided a Teflon-like surface to make it possible to gradually slide the structure across until it reached its final position.

One unusual feature of this particular bridge slide was that the jacking mechanism was able to work backwards as well as forwards. Dave Lawson, Network Rail's project manager for Crossrail west, described the approach as giving the bridge a reverse gear. This meant any incremental slide was not permanent – the structure could be jacked back to be absolutely sure that everything was proceeding according to plan.

Another challenge for engineers was to keep the structure at the right level to go on top of – rather than crash into – the central and east piers as the bridge launch progressed. To achieve this a 30 metre temporary 'nose' was attached to the main structure which guided the flyover across the piers. Once the flyover had reached its final position, this lightweight structure could be removed.

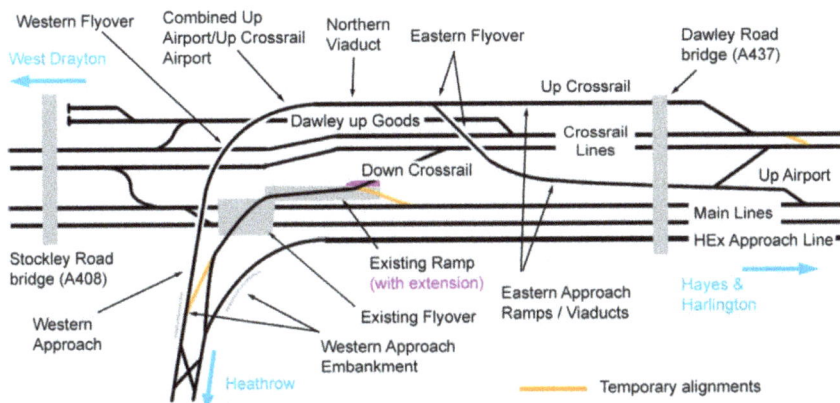

An expanded Stockley flyover routes Elizabeth line trains to
and from Heathrow above the Great Western main lines.

The new western flyover carries the Up Airport line. Trains from
Heathrow cross the Great Western corridor to join the Crossrail tracks
towards Paddington and central London. The new bridge saw action
well in advance of the opening of the Elizabeth line with use by
Heathrow Express services following commissioning at Christmas
2014. When the western flyover opened airport trains could be rerouted
across the new structure to allow work to expand and adapt the old
flyover and complete the final Elizabeth line-ready track layout.

By 2015 the last concrete beam to support the construction of a
second ramp at the Stockley flyover had been lifted into place.
A total of 146 beams – each weighing about 40 tonnes – were
installed to support the second ramp.

Christmas 2016 works included the opening of the second ramp
of the expanded Stockley flyover structure at Airport Junction.
The completed layout at Stockley means Elizabeth line trains
travelling in each direction on the Heathrow branch can leave/join
the Great Western corridor without delaying or obstructing
train movements on other lines. Following completion of the
Elizabeth line around 750 trains a week are able to regularly use
the flyover.

## Electrification

As part of the Crossrail programme it was initially envisaged that 12 miles of railway between Maidenhead and Heathrow junction, where the wires previously ended, would need to be electrified. As well as allowing the new Elizabeth line train fleet to run this would also support train operator GWR's programme of introducing electric trains on Thames Valley routes.

In September 2016 GWR launched a new service between Hayes & Harlington and London Paddington using Class 387 Electrostars. Following Christmas engineering works, the opening of a new bay platform at Hayes & Harlington on 3 January allowed GWR to increase the frequency of this service to half-hourly. The Hayes & Harlington-Paddington service replaced direct trains between the Greenford branch line and London Paddington.

By May 2017 electrification of the Great Western route between the Stockley junction and Maidenhead had been completed and GWR was able to extend its recently introduced electric trains running between Paddington and Hayes & Harlington to Maidenhead with the May national rail timetable change.

The Crossrail western outer electrification contract was awarded in October 2013 to Balfour Beatty after the company won the track infrastructure contract covering the same stretch of the Elizabeth line route.

## Reading

Crossrail (as legislated for in 2008) was to run trains as far west as Maidenhead. But on 27 March 2014 Crossrail sponsors Transport for London and the Department for Transport announced that Crossrail would be extended west from Maidenhead to Twyford and Reading. This would require further electrification but this was already planned as part of Network Rail's programme to electrify the Great Western route to Bristol and Cardiff.

Getting Crossrail to Reading had been much discussed over the years with questions as to whether suburban Maidenhead really was the most suitable place for the Elizabeth line's western terminus. In contrast Reading station had benefited from a major upgrade costing the best part of £1 billion and was only two stops west, offering a modern interchange with connections to Wales and the west, south west, Midlands and beyond.

Since 2009, when formal safeguarding directions were issued by government to protect against any developments that would prevent Crossrail operating between Maidenhead and Reading, much had happened to build the case for activating the extension. Remodelled Reading had gained five new platforms with capacity for a future Crossrail service factored into the design. The Great Western main line was in the process of being electrified all the way to Wales. So in infrastructure terms there were no major impediments – there would even be cost savings from reduced requirements at Maidenhead and Slough.

But, according to a letter written by rail minister Stephen Hammond to Reading East MP Rob Wilson, it was operational concerns that cemented the case for building the extension. Reading previously enjoyed four stopper services (not all stations) to and from Paddington and this level would be maintained under Crossrail. The only difference was that two of the four trains would be run by the Crossrail concession (by extending two of the four previously due to terminate at Maidenhead) with the remainder the responsibility of the Great Western route operator.

Had the Elizabeth line not been extended to Reading, the plan was to run two of the four Reading stoppers as a shuttle linking Reading and Slough. This would have overlapped the Crossrail service to Maidenhead, maintaining a four trains an hour stopper link to Paddington but requiring those using the shuttle trains to change (on to Elizabeth line services) to reach the London terminus.

For passengers from Reading and Twyford the need to switch trains to reach stations between Slough and Paddington was one drawback of this proposal. But for the railway the overlap of Crossrail and the Reading-Slough shuttle was seen as more significant with conflicting train movements presenting a threat to service reliability. The shuttle also required infrastructure changes at Slough.

Extending to Reading meant the shuttle idea could be dropped, the previous direct services to Paddington from Reading and Twyford could be maintained, fewer trains would be required and some planned infrastructure works could be scrapped. Stripping out risky overlapping services from the timetable provided flexibility for the future with Hammond citing the Western Rail Access to Heathrow scheme, pencilled in for completion by 2021 (but subsequently shelved indefinitely). On top of this, under the revised plan, two of the four stoppers from Reading and Twyford would offer direct trains to the West End, the City, Canary Wharf and beyond.

While extending Crossrail from Reading and Twyford (where the Henley branch line joins the Great Western corridor) unlocked new journey opportunities, a question mark remained regarding how popular it would prove with passengers. Reading also offers a high frequency, non-stop train service to Paddington and by December 2019 Hitachi trains introduced under the Intercity Express Programme were able to complete this journey in around 25 minutes. The Elizabeth line has removed the need to change trains to reach central London and beyond but, even after electrification and with trains skipping the lesser used stations, it still typically takes more than 50 minutes to travel from Reading to Paddington. Some passengers will find compensation in not needing to trek from main line platforms to Underground platforms at Paddington. But for others the convenience of a direct train will be outweighed by the frustration of sitting through multiple station stops.

## South east (Plumstead Portal to Abbey Wood)

The overground section of railway from the Plumstead Portal, where new infrastructure joins the existing North Kent lines, to the Elizabeth line terminus at Abbey Wood, is relatively short but has received plenty of attention thanks to Crossrail.

Major works involved adding two tracks to create a four-track rail corridor as far as Abbey Wood. This has been designed so as not to preclude a future extension along the safeguarded route to Ebbsfleet (see chapter 15). While the existing track was third rail electrified the new lines have overhead catenary – trains on all parts of the Crossrail network have power fed to them via overhead line electrification.

Also specified was a major upgrade of Abbey Wood station. In October 2010 Network Rail awarded Balfour Beatty the design contract, covering construction of a replacement station and four tracking the two mile stretch of railway from Plumstead. Enabling works subsequently involved building up embankments and taking possession of small pieces of land from adjoining residential properties that were required to create space to widen the railway. In 2013 Network Rail confirmed that Balfour Beatty would carry out the main works to upgrade the rail corridor from the new Plumstead Portal to Abbey Wood under a contract valued at approximately £130 million.

Crossrail's two new tracks were originally to be laid between the existing two lines (to/from London Bridge and Charing Cross) with island platforms built at Abbey Wood to allow step free access between Crossrail and Southeastern trains. However, a decision was subsequently taken for the two new tracks to sit beside the North Kent lines; this was driven by a change to the track layout on the approach to Abbey Wood in order to accommodate Crossrail Ltd's decision to specify new sidings at Plumstead.

This rethink emerged in 2013. At a meeting on 27 March Transport for London's Board was asked to give Crossrail Ltd permission to apply for a Transport and Works Act order in July that would allow construction and operation of a Plumstead depot for trains and infrastructure maintenance facilities.

Under the 2008 Crossrail Act, trains and infrastructure maintenance for the Elizabeth line were to be based at the new Old Oak Common depot in west London. Land by the Plumstead Portal was originally intended for use only as a temporary railhead to support delivery activities associated with the fit-out of the central Crossrail tunnels.

Closer consideration of Crossrail operations led to the conclusion that moving empty Elizabeth line trains back from Abbey Wood to Old Oak after the end of service each night was perhaps not the best use of resources. If units could spend the night at Plumstead, it would not only cut the time it took to get empty trains to their overnight stabling facility, it would also increase the time available for overnight maintenance of the central Crossrail infrastructure.

Existing Network Rail plant depot locations at Woking, Colchester, Hither Green and Reading were considered as possible alternatives to stabling trains at Plumstead. However, these locations would require both passenger and engineering trains to run empty across Network Rail tracks, which could have implications for timetabling as well as operating costs. Therefore, following the granting of a Transport & Works Act order, the temporary railhead was converted into a permanent train stabling and railway maintenance facility. Although overground the Plumstead sidings were delivered by Crossrail rather than Network Rail.

## Abbey Wood

On 8 July 2013 Crossrail Ltd submitted planning applications for the Abbey Wood station development to the London Borough of Bexley and the Royal Borough of Greenwich.

Described as the single biggest addition to the local rail network since the North Kent line was built in 1849, a new landmark concourse/ticket hall – replacing the 1980s-built ticket office with skillion roof – would be built above two new Crossrail tracks and those used by North Kent services. The scale of development was part predicated on a Crossrail Ltd pre-pandemic forecast that use of the station would nearly treble in the 15 years up to 2026. During the build temporary station facilities were installed in the old station car park.

Network Rail oversaw major redevelopment of Abbey Wood.

Designed by architect Fereday Pollard, the concourse of the rebuilt Abbey Wood station leads on to a public piazza next to the elevated Harrow Manorway dual carriageway, which carries traffic over the railway to Kent. Below are the two island platforms (four platform faces) with six lifts providing step free access to each platform.

The landmark feature of Abbey Wood station today is the roof above the 1,500 square metre concourse. Variously described as a manta ray and a Pringle, Austrian timber construction firm Weihag installed the glue-laminated timber wooden panels following its work on the Canary Wharf station park roof. The Abbey Wood structure is made up of 31 tonnes of steel beams and girders with the main timber beams each 45 metres long. It is finished with a zinc covering.

Drawings in the 2013 planning application showed a canopy over the Southeastern platforms but no canopy for Crossrail platforms. The official explanation pointed out that turn up and go service frequencies combined with trains waiting at the Elizabeth line terminus meant passengers would be unlikely to be exposed to the

159

elements. The local council's planning committee was not convinced and persuaded Crossrail to provide the missing canopies.

By the end of 2015 more than 100 piled foundations had been sunk to support the new station building. The following year concrete was poured for the new island platform at Abbey Wood station and by October the second of four new platforms at Abbey Wood had opened as part of the station redevelopment.

Commissioning the new Kent-bound platform (number 2) marked the completion of work to realign the tracks and rebuild platforms used by Southeastern services. In total 4.5km of new track was installed as part of the work affecting the North Kent line. The new London-bound platform (number 1) opened in February 2016.

Abbey Wood's new concourse opened in October 2017, providing access to Southeastern services in advance of the Elizabeth line opening. The temporary, modular building, housing ticket office and automatic gates during the station overhaul, was acquired by train operator GWR with the intention of redeploying it at Hanborough to enhance facilities at the Cotswold line station.[1]

## North east (Pudding Mill Lane Portal to Shenfield)

From the Pudding Mill Lane Portal to Shenfield, the north eastern branch of Crossrail was the easiest overground section of Network Rail's Crossrail commissions – no major structural works or station rebuilds[2] were required here.

But station enhancements along this stretch were specified even though the scale of these has fluctuated during the development of the new railway. Modifications were also needed to signalling, track and the existing overhead line electrification. Highlights of the north eastern branch investment included development of a new Ilford depot and building a freight passing loop at Chadwell Heath.

---

[1] Abbey Wood Crossrail ticket office heading west, Transport Briefing, 25 January 2018
[2] During development of Crossrail plans were drawn up for extensive upgrades at Romford and Ilford but these were subsequently scaled back

The most complex work required was to make the connection between the new Elizabeth line tracks, as they emerge from the Pudding Mill Lane Portal, and the existing Great Eastern Main Line. This is a significant railway junction which, as explained in chapter 4, involves Crossrail lines going under the new Docklands Light Railway viaduct to join the Great Eastern close to the site of the old Pudding Mill Lane DLR station.

To try to make best use of resources plans were altered to delay the implementation of works on this less challenging part of the overground route, rather than deliver the work concurrently with the western and Abbey Wood sections. This carried the risk that by deprioritising work the schedule could slip; indeed station upgrades remained incomplete when the Elizabeth line opened.[3]

By March 2014 Network Rail had awarded Costain the last remaining contract for major Crossrail overground works. With an initial value of £150 million, this included the design and build of station improvements at Romford and Ilford as well as enhancements to Forest Gate, Goodmayes, Harold Wood, Gidea Park, Chadwell Heath and Brentwood.

Platforms were extended to accommodate Class 345 trains and lifts installed at selected stations. At this stage agreement had yet to be reached to ensure step free access at Manor Park, Seven Kings and Maryland, the latter originally omitted from Crossrail plans because of short platforms.[4]

While Crossrail activity stepped up across the capital work on the Shenfield branch appeared to proceed at a more leisurely pace. It was

---

[3] Modern Railways, October 2022. In this issue I reported the 'significant completion' of all planned major station upgrade work by Network Rail for the Crossrail project in east and west London. This was after the (repeatedly delayed) Elizabeth line core opening in May 2022

[4] Crossrail bolts on another station to line, Transport Briefing, 14 August 2006. In 2006 Cross London Rail Links and the London Borough of Newham reached an agreement for trains to stop at Maryland using selective door opening.

Construction of new Docklands Light Railway viaduct with
former DLR Pudding Mill Lane stop seen in background.

not until early 2017 when Transport for London finished work
modernising Manor Park, Seven Kings and Ilford stations (additional
work would later be specified). Ticket hall and gateline refurbishments
at Brentwood and Chadwell were finished in the spring.

By June 2017 infrastructure works on the Great Eastern Main Line
had been completed ahead of the launch of the Elizabeth line service.
Work would, however, continue on improvements to track and
sidings. Shenfield, the north eastern Elizabeth line terminus, gained
a new platform (number 6) dedicated to the new service, and an
extended platform 5, as well as three new sidings, each capable of
accommodating Class 345 trains.

The entire railway junction at Shenfield, including track, signals and
overhead power lines, has been improved in order to allow services to
freely pass through the junction without having to wait for other trains,
reducing the risk of delays and ensuring the Elizabeth line timetable is
operationally robust. More than 5,000 metres of new track and 26 new
track switches were laid as part of the junction remodelling.

Although the scale of work required on the north eastern Elizabeth line branch was less than on the other overground sections figures show it remained substantial: Over 40 platforms across 13 stations were adapted to accommodate wider trains and lifts installed to ensure step free access to platforms. More than 200 CCTV cameras were fitted between Maryland and Brentwood to allow train drivers to view the entire length of the platforms. New sidings have been installed at the Ilford depot.

## Step free access for all

While the 2008 Crossrail Act specified upgrades to many existing stations on the parts of the Elizabeth line network that run overground, some smaller, less-frequented stations were to be left with few improvements. This raised the prospect of a multi-billion pound showcase rail project which, even when complete, would have some stations that were inaccessible to those in wheelchairs or with pushchairs.

Accessibility campaigners successfully challenged this and in October 2014 Transport for London said it would provide £19 million to ensure step free access was provided at Seven Kings, Maryland, Manor Park and Hanwell – stations where accessibility improvements were not funded as part of the original Crossrail budget. A month later the Department for Transport announced that it would make available £14 million so that Langley, Taplow and Iver stations could also be fully accessible. These two funding commitments meant that all stations on the Crossrail network would have step free access from street level to platforms.

As well as funding step free station improvements TfL also decided to pay for extra enhancements at stations where it would be responsible for the station facilities in addition to providing the Elizabeth line train service. In 2015 it put the £33 million agreed for access improvements together with extra money to create the Crossrail On Network Stations Improvement Programme – or ONSIP. This £93.6 million project consisted of two strands – installing a total of 18 lifts at the seven stations mentioned above and providing additional improvements, not due to be delivered as part of Crossrail but desirable to achieve a consistency with the TfL Overground network standards, at 22 surface stations.

The £33 million accessibility improvements specified were:

Hanwell – one lift installed internally and one within a new, free-standing structure

Iver – two new lifts attached to existing footbridge plus a ramp to the fast line platform

Langley – a new footbridge with three lifts

Maryland – three new lifts installed inside the existing station buildings

Manor Park – a new footbridge with three lifts

Seven Kings – a new footbridge with three lifts

Taplow – a new footbridge with two lifts

The remaining £60 million of ONSIP focused on delivering London Overground-style improvements. This was partly driven by a desire to ensure all stations had ticket gates – and were therefore not vulnerable to fraudulent travel – and partly to provide larger public areas at stations to enhance the passenger experience. TfL also decided to repair buildings and other structures which had been neglected over many years. With TfL taking on 125 year leases for Elizabeth line stations it decided there was little point in deferring work.

Improvements varied from station to station but included building renovation, decluttering, redecoration, new lighting and improved seating. Crossrail train operator MTR was charged with managing the works and let contracts for two to three stations at a time.

While Crossrail funded a new concourse at Ealing Broadway ONSIP paid for remodelling the lower level of the station to improve the interchange between Elizabeth line and London Underground services. At West Drayton, Burnham and Iver new entrances were built to permit ticket gates to be installed and, in the cases of Iver and Burnham, to create an adequate station approach and entrance.

# 7. THE TRAINS

Plans for the Elizabeth line train fleet were revealed in 2010 as details were unveiled of what would become the single largest procurement for the Crossrail programme. This, despite the trains not figuring in the headline Crossrail cost figure.

A prior information notice appeared in the Official Journal of the European Union at the end of July and was soon followed by a contract notice seeking expressions of interest from train manufacturers by the end of January 2011.

Crossrail Ltd said it intended to buy around 60 trains. It wanted sufficient vehicles to operate 57 diagrams with an option to provide an additional 15 – potentially 72 in total. Trains would have to be around 200 metres long and were expected to have 10 carriages with capacity for up to 1,500 passengers in total. However, a limited number of diagrams would use 160 metre, 8-car formations when Crossrail services began.

In a statement to parliament in November 2010 Transport Secretary Philip Hammond promised that 600 carriages – 60 10-car trains – would be bought for Crossrail. This would be a net gain of 400 vehicles after taking into account the existing local services operating from Paddington and Liverpool Street that Crossrail was set to replace when through London trains commenced.

The OJEU notice put the capital value of the rolling stock contract – referred to as X2234 – at between £1 billion and £1.9 billion for the trains and a new depot. Crossrail Ltd specified that the trains should have a design life of 35 years and with the package covering not only the build but maintenance of the vehicles, the chosen supplier stood to derive a long-term return from Crossrail stretching into the 2050s. Annual maintenance costs were estimated at between £25 million and £50 million a year at 2010 prices.

As well as building and maintaining the trains the successful bidder would also be expected to design and build a train care depot to house the new fleet. Crossrail Ltd would provide a site at Old Oak Common in north west London for this purpose.

An early principle behind the train procurement was to buy tried and tested technology. Investing in kit that could be shown to have worked elsewhere was seen as more important than having the latest offering from manufacturers. "We aim to benefit from, and build upon, the existing capabilities of the rolling stock industry rather than requiring a wholly new concept or design," said Rob Holden, Crossrail Ltd chief executive between April 2009 and July 2011.

The tender documentation for Crossrail rolling stock specified a 350 tonne upper limit on the unladen weight of a 200 metre long 10-car train formation. This was lighter than recently delivered equivalent electric rolling stock classes – those that achieved the most up to date crash-worthiness standards – while remaining deliverable with existing proven technologies.

Bidders to supply the new trains would be expected to demonstrate that on a standard network journey a 200 metre long set would operate with an energy efficiency of 24 kWh per train kilometre – equivalent to 0.16 KWh per passenger kilometre and 55g of carbon dioxide per passenger kilometre. Crossrail trains were to have regenerative braking, on board energy metering and intelligent heating and ventilation systems.

One other consideration for potential bidders was the request from Crossrail Ltd to put in a priced option for additional power supply equipment that would allow Crossrail trains to run on third rail fitted routes. This would allow a comparison to be made between the cost of erecting overhead wires on the Abbey Wood branch and buying trains capable of using existing power supplies south of the River Thames.

In March 2011 Crossrail Ltd shortlisted five train manufacturers to build the new fleet, which by now had been officially registered as Class 345, in contrast to the Class 341 previously designated for rolling stock for the abandoned 1990s CrossRail scheme. France's Alstom, Canada's Bombardier, Japan's Hitachi and Germany's

Siemens would go through to the next stage of the competition along with Spain's Construcciones y Auxiliar de Ferrocarriles (CAF).

In coming to a decision Crossrail would not only have to work out who could offer the best trains for the new railway but also navigate a political minefield. With recent mega rolling stock contracts for Thameslink and the Intercity Express Programme having gone to Siemens and Hitachi respectively, the pressure was on for Bombardier to secure a contract to keep its Derby Litchurch Lane train building centre in business. Hitachi, having committed more than £80 million[1] to establish a UK rolling stock facility in Newton Aycliffe, County Durham to help build the Intercity Express fleet, was keen to secure another order to strengthen the viability of its investment in the north east.

The Crossrail procurement process was to have two rounds. In the first bidders would provide technical proposals and set out their approach to securing the finance necessary to deliver the new trains and Old Oak Common depot. At least one bidder would be eliminated from the process before those remaining were invited to participate in a second round, which would focus on bidders providing fully funded proposals. At the end of round two a preferred bidder would be selected.

## Financing the fleet

The government's initial plan was to use a private finance initiative to cover the cost of the Crossrail trains and depot. This had the advantage of avoiding an upfront cash commitment of around £1 billion which would increase public spending, show up on government accounts and thus reduce financial room to manoeuvre for HM Treasury.

The PFI approach would see the Crossrail rolling stock and depot procured as a service concession – complete with private finance of

---

[1] Hitachi inks contract to construct IEP factory, Transport Briefing, 14 May 2013

the capital cost to be paid for over time through access charges paid by the operator of Crossrail services. This in turn would come out of ticket revenue collected once Elizabeth line services began. Passengers would get new trains and the actual cost would be hidden from all but the most forensic of accountants.

A similar approach had been used to finance the Class 700 Siemens Desiro City Thameslink and Class 800/801 Hitachi Intercity Express fleets. Yet using financiers and multiple institutions to bankroll these trains had proved complicated; not only had the procurements taken longer than expected but there were concerns that financing had played too great a role in the assessment criteria and acted as a distraction from the train specification.

One attractive feature of the PFI approach for government was that it offloaded the risk of things going wrong with the procurement on to the private sector. But how much risk did the Crossrail train order really represent? Set aside the challenges of integrating multiple train control and signalling systems and the new trains would be similar to those that had been ordered before. At a fundamental level a train has wheels and carries passengers; the shortlisted bidders would be able to deliver just such vehicles and, although their latest product offerings would naturally feature, the train manufacturers were expected to offer pretty much the same as they had tendered for the Thameslink order.

There was, however, one key difference between the new Thameslink trains and the Crossrail train spec – doors. Thameslink Class 700s stick with the two double doors per carriage (on each side) which is common for suburban train fleets across Britain. But Crossrail trains were to have three sets on each side, a feature seen on metro and underground railways around the world. This is not a technicality, it is vital to reduce dwell times at the busiest stations on a route.

I witnessed the problem five days a week as a commuter using Thameslink services in the 2000s, prior to the introduction of the Siemens' trains. At most stations two sets of doors were perfectly adequate. But through the central London core stations – Blackfriars,

City, Farringdon, King's Cross St Pancras – services were delayed every day because of the time it took people to get on and off each train. From the platform you could see passengers wishing to alight queuing in the space between seats all the way down a carriage. They would then have to get through a narrow vestibule area in which other people would be standing, and luggage for those travelling to Luton or Gatwick Airport would be stored, before eventually disembarking. Meanwhile, those on the platform would wait anxiously and impatiently, positioning themselves strategically to avoid blocking the flow too much while having the chance to scramble for one of the few seats available in order to avoid standing up for the journey home.

This was a mess and full marks to the Crossrail team for being determined to engineer out any possibility of repeating this chaos.[2] However, having three sets of doors rather than two would make the Class 345 fleet a custom order and limit opportunities to move it elsewhere in the future, redeployment of trains being a common feature of Britain's privatised railway. If Crossrail trains could only be used in London financiers would have fewer options for the fleet and would be beholden to a single operating authority.

But having watched previous PFI train procurements that operating authority – Crossrail Ltd parent Transport for London – was keen to avoid a repeat. It pushed government to rethink and warned that a similar financing delay to that which occurred with Thameslink and IEP risked the potentially embarrassing scenario of Crossrail infrastructure being completed with no trains to run on it. TfL also had recent experience of buying trains outright from its London Overground Class 378 Capitalstar procurement and saw no reason why this perfectly good model should not be replicated for Crossrail.

---

[2] Although the Class 700 units retain the conventional two sets of doors arrangement the layout of the carriages includes wider gangways and more circulation space around the vestibules than the Class 319s they replaced

On 28 February 2012 Transport Secretary Justine Greening accepted the argument. In a statement to parliament the Secretary of State said: "this [Crossrail rolling stock and depot] contract will provide a significant element of public investment, alongside private finance, optimising the balance of public debt and transfer of risk to the private sector. This approach will help ease the costs of debt repayments to the public purse, as well as reduce bidders' requirements to raise debt and equity, while still transferring significant risk to the private sector ensuring that we secure value for money. First and foremost, the successful bidder must be able to deliver the right trains and depot facilities."[3]

According to Transport for London, public funding worth £350 million, representing 35% of the estimated £1 billion capital cost of Crossrail rolling stock and depot facilities, had been agreed by the government.

The transport secretary's statement continued: "the Invitation to Negotiate includes requirements for 'responsible procurement'. This means that bidders are required to set out how they will engage with the wider supply chain and provide opportunities for training, apprenticeships, and small and medium size businesses within their procurement strategy. Bidders are also required to establish an appropriate local presence to manage the delivery of the contract... bidders are being asked, in the Invitation to Negotiate, to specify from where each element of the contract will be sourced. This is not an assessment criterion in the decision process however the successful bidder will be required to report against their proposed estimates."

## Thameslink impact

This change of tack was an acknowledgement from government that the Thameslink procurement had missed a trick. The period between June 2011 when Siemens was named preferred bidder to supply 115 trains and June 2013 when the German company sealed the

---

[3] Justine Greening, written statement to parliament, 28 February 2012

£1.6 billion order was accompanied by an outcry from trade unions and others angered that the Thameslink trains would be built outside the UK. Siemens was left to issue annotated diagrams to the press showing that at least some of the train components would be manufactured in the UK.

While it was too late to rethink the Thameslink procurement the government signalled that public sector led train orders would in future no longer be decided solely on trains and price. Crossrail bidders would need to provide details of where jobs would be created and how UK supply chains would benefit. As Justine Greening's statement made clear, whoever succeeded in securing the order would need to give ministers evidence they could share showing how the deal supported British jobs and businesses.

By March 2013 the government and Transport for London had confirmed that the Crossrail trains and depot would be conventionally procured following concerns that a privately financed deal would not guarantee trains were ready in time. TfL said Crossrail Ltd managed the procurement process in a way that allowed the private financing element to be stripped out.

So where would TfL get the money to buy the trains? Answer, Europe and Canada. In December 2013 the European Investment Bank agreed to lend Transport for London £500 million towards what was now a £1.2 billion procurement. The EIB pointed out that it was no stranger to providing finance for UK rail projects having previously stumped up £6.8 billion to support schemes including Thameslink, High Speed 1, Manchester Metrolink and London's DLR and Overground networks.

In early 2015, after the contract for Elizabeth line trains was awarded, Transport for London's Finance and Policy Committee approved a deal with Canada's export credit agency EDC, wholly owned by the government of Canada, to lend TfL a further £500 million for Crossrail rolling stock and the Old Oak Common depot. Together with the EIB loan this gave TfL access to borrowing facilities that covered most of the cost of the train and depot procurement.

## Train supplier selected

With a PFI model ruled out, and therefore no prospect of reusing its Thameslink order finance partners, but, more importantly, having seen the renewed interest in building trains in Britain that was now being baked into Crossrail, Siemens took the pragmatic decision to shift its attention to future orders and withdrew from the Crossrail competition.

Bombardier landed the Crossrail train contract, announced on 6 February 2014. It was a win for Derby where the company's historic Litchurch Lane site would assemble the new fleet. Litchurch Lane was the only surviving train building production line in the UK but was facing competition from Hitachi's recent Newton Aycliffe investment. Bombardier had warned that the Derby facility faced closure if it failed to secure the Crossrail deal.

An important point to note is that the contract for the trains was awarded by Transport for London, not Crossrail Ltd. With the latter being an operating unit of TfL this might seem insignificant but the organisational split between procurement of Crossrail trains and Crossrail infrastructure would have ramifications as the programme progressed.

Later that month Bombardier valued the order at approximately £1.3 billion with this figure including the supply of 65 9-car trains and maintenance for up to 32 years. The contract included an option for 18 additional trains and also covered construction of the new train depot at Old Oak Common.

In terms of the benefits for UK businesses the train contract promised to support 760 UK manufacturing jobs plus 80 apprenticeships. The construction of the maintenance depot at Old Oak Common would involve 244 jobs plus 16 apprenticeships. When fully operational the depot would support 80 jobs to maintain the new fleet of trains. An estimated 74% of the train contract spend would remain in the UK economy.

Trains for the Elizabeth line would use Bombardier's Aventra platform. They would be over 200 metres long and – despite being nine rather than 10 cars – would provide capacity for up to 1,500 passengers. The trains would have air-conditioning and interconnecting walk-through carriages; there would be an emphasis on energy efficiency and the use of intelligent on-train energy management systems; they would include systems for condition-based maintenance including Bombardier's Orbita predictive maintenance capability.

With a focus on environmental credentials and minimising track wear the new rolling stock would be 25% lighter than existing trains on the rail network of an equivalent length. Having been set a target of designing a train weighing less than 350 tonnes, the total train mass achieved weighs in at around 319 tonnes.

There are some smart systems on board to maximise energy recovery during braking and minimise waste in managing the on-board heating and cooling systems for passengers. LED lighting features automatic dimming and there's an intelligent stabling system to minimise energy use when the trains are parked. An advisory system tells drivers to moderate power consumption if services are running ahead of schedule.

## Getting ready for Aventra

Despite being an established train building site Litchurch Lane needed new investment in order to be ready to run up a Crossrail fleet. Bombardier got preparations underway with the award of a £12.5 million contract to Balfour Beatty to create a production, testing and office facility for the new Crossrail trains at its Derby rolling stock centre.

The development at the Litchurch Lane site measured approximately 250 by 40 metres, enabling it to handle four 10-car trains – longer than the 205 metre 9-car Aventra sets on order. Work covered by the contract included the construction of four multi-functional train lines, each with full length overhead lines and inspection pits containing a range of services that would be fully accessible by rail at both ends, connecting into existing test track facilities.

This contract award in April 2014 came just a couple of months after Bombardier opened a £1.6 million test facility for the development of Crossrail trains at Derby. The 'iron bird' integration facility was inspired by the aviation industry, where advanced pre-production testing takes place in a simulated environment. The intention was to identify and resolve any technical problems with the new rolling stock within a factory setting rather than out on the track.

July 2014 saw Bombardier agree a £142 million contract with Vinci Construction's Taylor Woodrow arm to build the main Crossrail train depot at Old Oak Common in west London. The four year project involved redeveloping the railway land at Old Oak to provide office accommodation and train maintenance and stabling facilities. Today the operations, maintenance and control building contains nine maintenance tracks which allow a full train to be lifted and have several pits providing options for accessing the underside of rolling stock in order to carry out repairs. Stabling areas provide a place for the trains to be kept and for cleaning to take place while trains are not in use. The first Class 345 arrived at Old Oak Common in late 2017.

An upgraded former Abellio Greater Anglia depot at Ilford also helps manage the Elizabeth line fleet. The site is located next to the Great Eastern line between Ilford and Seven Kings and covers an area of approximately 12 hectares. Modifications to the railway and overhead line electrification created stabling berths for Class 345 trains. Investment was also made in staff accommodation, access roads and a signalling control building.

## First look at Elizabeth line fleet

By the end of 2015 Bombardier had completed production of a Crossrail train carriage prototype. This first bodyshell, assembled at Litchurch Lane, would be used to refine the design and the manufacturing techniques needed for the full production of what by now was a 66 train order (594 carriages) in order to accommodate the now confirmed extension of the Elizabeth line from Maidenhead to Reading.[4]

---

[4] In March 2017 the order was further increased to 70 trains to accommodate an increase in Elizabeth line train frequencies

The unpainted bodyshell was followed by publication of new computer generated visuals of the finished train. Given that the media had had to make do with two generic 'blue train' and 'purple train' pictures since the Crossrail Act secured Royal Assent in 2008 there was considerable excitement generated by the release of a proper Elizabeth line Aventra mock-up. The CGI showed the Class 345 exterior would be two shades of grey with a black front end; this sparked much discussion given that it dispensed with the yellow cab ends customary on train fleets in service on the national network at this time. Crossrail purple survived in Elizabeth line roundels and two-tone strips along carriages.

Not only was there a new exterior picture to examine, Transport for London also released shots showing what the interior of the new trains would look like. The promotional material explained that dark floors and natural finish materials would ensure the trains retained their high quality feel for years to come. Light coloured ceilings were designed to maximise the feeling of height and openness inside the vehicles.

As one would expect for a modern commuter railway, Crossrail specified clear areas around the three sets of doors in each carriage to allow quick and easy boarding and alighting. Seating was described as a mixture of metro-style and bay – which meant partly longitudinal with limited 2 x 2 chairs. Trains would be driver-operated with on-train customer information systems delivering real-time travel information, allowing passengers to plan their onward journeys while onboard. There would be four wheelchair spaces on each train plus a number of multi-use spaces available where seating could be tipped up to accommodate prams or luggage.

At more than 204 metres in length, the Aventras are over 50% longer than the longest London Underground trains currently in service. Featuring intelligent lighting and temperature control systems, the trains have regenerative braking, returning electricity to the power supply. Transport for London and train manufacturer Bombardier worked on the design of the new trains with Barber & Osgerby acting as design adviser for the project.

What the trains don't have – and what Thameslink Desiro City trains do – is toilets. For end-to-end Elizabeth line journeys this could be a cause of discomfort although many people travelling the full extent of the western route choose to pick up faster, non-stopping services. Not having toilets on board does allow more seats to be fitted into carriages and does away with the need to clean and empty waste tanks. The decision not to put toilets on the trains attracted media attention; Transport for London's response was to state that at least three quarters of Elizabeth line stations were expected to have public toilets by the time the new service launched.

Class 345 Elizabeth line trains feature open carriage
connections which enhance the sense of space
and give passengers clear sight lines.

## Bring on the build

These pictures signalled the start of fleet construction in earnest. Bombardier now had a good two years plus before the first Class 345 was scheduled to enter passenger service in May 2017. The initial sets to be delivered would be seven carriages long and were set to operate between Liverpool Street and Shenfield.

Testing of the new trains took place on a short test track at Litchurch Lane and the more extensive testing facilities at nearby Old Dalby,[5] close to Melton Mowbray, before coming to London for driver training and further tests in east London.

In early 2017 a Class 345 Aventra built for the Elizabeth line completed intense weather testing to ensure the new fleet could handle whatever climatic conditions it encountered once passenger services began. Train builder Bombardier delivered the front carriage of a test train to the Rail Tec Arsenal test centre in Austria for three weeks of testing in temperatures ranging from -25°C up to +40°C. Opened in 2003, the facility in Vienna was the only place in Europe that could carry out this level of testing on trains.

The heating, ventilation and air-conditioning system, windscreen wipers and demister, train horns and the traction motors were put through intense simulations of hot, cold, windy and foggy weather to check that they could still function in whatever weather conditions the train might face when operating through London and beyond.

## Crossrail's main train depot

The depot at Old Oak Common was designed to house and maintain up to 42 of the Elizabeth line's 70 new trains at any one time. It employs more than 100 staff and has 33 stabling roads with nine roads for the heavy maintenance of wheels, motors and other rail components. Two train washes allow trains to be deep-cleaned regularly. Bombardier was contracted to operate the facility as part of the 32 year-long construction and maintenance contract that was signed in 2014.

---

[5] At the time of writing the Old Dalby facility is officially referred to by owner Network Rail as the Rail Innovation and Development Centre

As part of the depot design and build deal Taylor Woodrow developed a hybrid renewable energy scheme to reduce carbon dioxide emissions and operational costs. The design features site-wide LED lighting, solar thermal, solar photovoltaics, ground source heat pumps and combined heating and power units, linked with a thermal energy store as part of a system that can switch between energy sources as demand and availability dictate. Overall the integrated technologies are predicted to have reduced operational carbon emissions from the depot by 35% compared to had they not been adopted.

Innovative technology also features in the automatic vehicle inspection system used at the depot. This is designed to increase train reliability by scanning and analysing trains as they enter the depot every 48 hours, reducing the overall time needed for maintenance.

The Elizabeth line's main train depot is located on the site of a mega regeneration scheme which will be anchored by the High Speed 2 Old Oak Common station. Unfortunately the depot now sits on prime development land and will be rather in the way. Options being considered include building a vast deck that would allow property to be built above it or demolishing and relocating the depot.

## Sale and leaseback

Having eventually secured permission from government to take charge of buying trains for Crossrail, in March 2019 Transport for London finalised an agreement to sell the Class 345 Elizabeth line fleet.

Under a sale and leaseback deal the rolling stock was sold by TfL to 345 Rail Leasing, a consortium made up of Equitix Investment Management, NatWest and SMBC Leasing. Selling the trains generated a return of nearly £1 billion for TfL which it is using to buy new trains for the London Underground Piccadilly line.

In the UK most train fleets are owned by rolling stock leasing companies which lease sets to passenger and freight train operators. TfL has used similar leasing arrangements to introduce new trains on to London Overground routes since it began running services in November 2007.

Because the Class 345 fleet is built for a bespoke underground line the trains cannot be used on other routes and therefore the sale was conditional on TfL leasing the units over a 20 year period. At the end of the term it has an option to buy back the units.

## The return of Alstom

French rolling stock manufacturer Alstom was bullish about its prospects for securing both Thameslink and Crossrail train orders. In September 2009 I attended a launch event in west London when the UK version of the company's X'Trapolis train design, based on trains already used in Australia, Chile and Spain and featuring technology from the company's high speed trains, was revealed.

Alstom really thought it was on to something. The X'Trapolis featured Bogie Offset Articulation technology which, in simple terms, meant trains had fewer wheelsets. The design reduced the number of train bogies required by up to 30% compared to existing rolling stock, allowing the carriages to be shorter and wider than vehicles currently in service. In turn this reduced the train weight by 28%, improving efficiency and environmental credentials as well as cutting down on regular maintenance requirements.

Inside, the shorter carriage lengths translated into shorter distances between doorways – a factor we have already seen is key to allowing passengers to embark and disembark quickly. Alstom said the wider carriage design also allowed it to provide broader gangways between seats.

Yet the innovation of the X'Trapolis failed to translate into UK train orders. Amid concerns during the Thameslink procurement that the articulated bogie mechanism could result in greater wear and tear on track, and following Crossrail's insistence, wherever possible, on tried and tested technology, Alstom realised it would have to jettison the Bogie Offset Articulation technology if it was to have any chance of pursuing the Crossrail train order. In August 2011 it emerged that Alstom had quit the competition to supply trains for Crossrail;

Alstom said it had become clear that the company did not have a suitable product for Crossrail without incurring substantial UK development costs.

But that's not the end of the Crossrail Alstom rolling stock story. On Thursday 22 June 2017 the first Bombardier Aventra Class 345 entered passenger service on the Liverpool Street (high level) to Shenfield surface route. Yet when Elizabeth line trains began running in 2022 the same Class 345s were being described as an Alstom fleet.

The reason? Alstom bought Bombardier's transportation assets in a deal that took effect on 29 December 2021. Selling up allowed Canada's Bombardier to reduce its debt while giving Bombardier shareholder Canadian investment and pension manager Caisse de dépôt et placement du Québec a substantial stake in Alstom. With the deal complete Bombardier intended to focus exclusively on the business aviation market. Elizabeth line trains remain the same but the name of the company responsible for making and maintaining them has changed.

Aventra trains, such as this one operating on the Elizabeth line's north eastern branch, are maintained by Alstom.

180

# 8. THE TECHNOLOGY

Looking back at the development of the Elizabeth line I am struck by how much this is a story of the interface between old-tech – trains travelling on metal rails – and emerging computer-based technology. For all the opportunities provided by new systems, a recurring theme in this chapter and the next will be how difficult these proved to unlock.

Crossrail is sometimes described as a digital railway; this essentially means that the different railway systems are controlled by computers and, invariably, interlinked. Crossrail Ltd estimated that the central section of the Elizabeth line has around half a million of these computer-controlled assets, such as fire safety systems, CCTV or platform screen doors. Each central station has 80 major systems.

Networked systems have many benefits but testing and installation can be challenging – change one asset and that may affect all the others that are connected. Still, computer-controlled, interlinked systems will hopefully benefit the railway in the years to come. Rail for London Infrastructure will use the digital assets to monitor the condition of the routeway and relevant stations and automatically create maintenance work plans, helping to predict what work is needed and when, reducing the need for physical inspection. Will this offer an improvement on a conventional 'analogue' railway? – time will tell. According to the National Audit Office 'there is no baseline or comparator to determine whether this system will reduce maintenance costs in future'.[1]

## Building Information Modelling

From early in the programme Crossrail championed the use of Building Information Modelling (BIM) – a process which brings together and shares digital information used by designers, architects, engineers and builders.

---

[1] Third National Audit Office Crossrail report: Crossrail – a progress update, 9 July 2021

The Modelling part of BIM will be recognisable to many people in the form of CADCAM and three-dimensional modelling software. 3D representations were widely used across the Crossrail programme but BIM goes beyond graphics by tying in extensive documentation, some of it auto-generated, which could help those building and ultimately operating the Elizabeth line to make better decisions.

During construction Crossrail licensed software from American technology firm Bentley and insisted that all its tier one contractors used it. The aim was to ensure all contractors had consistent access to detailed designs across the project; with multiple contracts in place at each station understanding the interfaces between these was essential.

The system operated like a library with contractors able to propose changes which could then be reviewed by Crossrail. Careful version control ensured there was only one version of live documents.

Although BIM was used extensively during construction its power is only now being unleashed; with the opening of the railway it became a platform for asset management. The Elizabeth line is made up of millions of assets and understanding these is crucial to the maintenance and management of the completed system. Crossrail has attempted to future-proof the structure of the information it has collected – remember, the iPhone had not been invented when the programme got underway – so that it can be accessed by those tasked with the Elizabeth line's upkeep. Information gathering is still very much active – in part through train-based monitoring equipment, station sensors and other smart systems as opposed to manually filling out forms.

BIM technology is widely used in the automotive and aerospace sectors but construction projects have been slower to embrace its potential. This may be part of the Elizabeth line legacy: prior to Crossrail no other transport programme in the UK had collected information all the way through construction with a view to it being used in the day-to-day operation of the live railway.

## Platform screen doors

The genesis of London's transport system, with lines dating back to the nineteenth century, means platform edge doors – standard issue with modern transit systems – have been something of a novelty. London got its first platform edge doors with the opening of the Jubilee line extension in 1999. Crossrail would be the city's next application of the technology, making it a fundamental part of millions more passenger journeys.

The term 'sliding doors', providing access between platform and train, fails to articulate the importance of this part of the Elizabeth line infrastructure. The doors are in fact one component of what is often referred to as platform edge screens, running from platform surface to tunnel roof and also including an upper service wall supporting lighting, communications, and cabling, plus a smoke-extract duct positioned over the track, as well as the structural frame on which all these elements are supported.

All the underground Elizabeth line stations have platform edge screens, installed to allow future extension from 9-car to 11-car trains, which means at each station more than half a kilometre of screens is required. If stacked they would be taller than the Leadenhall Building (also known as 'The Cheesegrater') in the City of London.[2]

There are multiple reasons why some sort of platform screening is now accepted as a requirement for new underground railways – preventing accidental falls on to tracks as well as the possibility of a deliberate pushing attempt, and stopping litter and dust particles entering the tunnel environment. For the Elizabeth line, the platform edge screens also have a major role to play in tunnel ventilation and the early decision to adopt the technology had far-reaching impacts on other parts of the project.

---

[2] Crossrail Learning Legacy, Engineering design of the platform edge screens for Crossrail mined stations

Typically underground station ventilation ducting – for day-to-day use and also smoke extract during a fire – runs the full length of a platform, but on the platform side of any doors that are installed. For Crossrail the decision was taken to place the ducting over the track.

This allows ventilation to be better managed, eliminating leakage through stations to such an extent that Crossrail decided it would not need six proposed ventilation shafts and head-houses in London. This decision to put the ducting on the track side of full height platform edge screens is said to have enabled a major reduction in the capital cost and land-take of the railway.[3]

Below the central London station platforms – facing the track – are square ducts which suck out brake dust and other pollutants from trains, keeping the Elizabeth line tunnel environment cleaner than the older Tube railway tunnels.

Underneath the platforms space is split 50/50 between the ducts for this extractor system and a services tunnel for cables and pipes. Above the railway tracks are smoke extract ducts which are unlikely to be in regular use but, if there were to be a fire underground, would swiftly evacuate any smoke – getting rid of the principal danger of an incident of this kind. In terms of normal ventilation underground, the movement of trains is the primary mechanism for keeping air in the tunnels fresh.

With no ducting required above platforms and all lighting, signage, public address and associated cabling located on the vertical face of the platform edge screen, above the screen door, ceilings could be higher, creating up to five metres of headroom and markedly improving station ambience compared to older underground stations in the capital. Light from the light boxes above the doors reflects off the tunnel cladding to create a soft, diffuse ambience without the harsh glare of overhead lights.

---

[3] Crossrail Learning Legacy, Engineering design of the platform edge screens for Crossrail mined stations

In 2014 Knorr-Bremse rail systems was selected to provide the platform edge screens for the Elizabeth line. The screens and accompanying software were designed by the company's Westinghouse division at Melksham in Wiltshire and tested at its Wolverton rail facility in Buckinghamshire.

Working out how to attach the platform screens required considerable head scratching with the use of sprayed concrete lining in the platform tunnels, mined from the running tunnel bores, not suited to suspending heavy screen frames. Supporting the screens from platform level was considered but given that, at the time of design, the specification for Elizabeth line trains – including where the doors would go – had not been finalised, there was a reluctance to drill down into the platforms when configurations could change. The solution ultimately chosen was to provide vertical and horizontal fixings into the tunnel crown, a mounting system that allowed for a certain amount of flexibility – not only for changes to the platform screen configuration but also station specifics such as the platform curvature at Tottenham Court Road.

## Auto-reverse

One of the most intriguing technologies employed on the Elizabeth line is the auto-reverse system, used to turn trains around at Westbourne Park, just beyond Paddington.

As explained below, in regular service Crossrail trains drive themselves through the central operating section despite having a nominal driver in the cab. But a train driving itself, stopping and then driving in the opposite direction – without a driver in the cab – that's sufficiently radical to attract attention.

The need for an auto-reverse system is prompted by the disparity between the Elizabeth line's eastern and western routes. To the east services split between the Shenfield and Abbey Wood branches. When they come together in the central section there can be 24 trains per hour heading west, a service frequency not justified – or currently possible – beyond Paddington.

So, some of these trains are turned around in the Westbourne Park sidings, west of Paddington, ready to provide the next eastbound departure. When a train arrives at Paddington from the east the driver will, on departure from the station, select 'auto-reverse'. This will then provide ample time for them to walk through the nine carriage train and seat themselves in the cab at the opposite end, ready to oversee the next eastbound departure from Paddington. While they are doing so the train at first continues west, without the driver in the cab, before stopping, restarting in the opposite direction and completing the journey to the Paddington eastbound platform.

As with most clever computer systems it took Elizabeth line engineers rather longer than expected to get this one working. By late 2021 a hold-up securing safety acceptance of the auto-reverse functionality contributed to delays in test running and accumulating the mileage required to evidence reliability. The Elizabeth line consequently opened in May 2022 without auto-reverse working. That meant that for trains terminating at Paddington drivers needed to continue west before manually stopping the service at Westbourne Park, walking through the carriages to the other end, and then restarting the train in the opposite direction.

When stage 5B launched in November 2022 auto-reverse still was not ready, limiting the frequency of trains that could run. To ensure the full Elizabeth line service pattern could operate the technical issues had to be resolved. This was finally achieved in time for the stage 5C opening in May 2023.

Now fully functional Crossrail's auto-reverse system supports the current peak service as well as any future expansion to 30 trains per hour. Although only in regular use at Paddington/Westbourne Park authorisation has been received and testing completed to allow use of the system at the Fisher Street crossover in the central tunnels. This may be required during future engineering work. The system also has a potential application between Abbey Wood and the Plumstead sidings, subject to regulatory approval.

## The promise of ETCS

For all the innovation across the Elizabeth line estate the most important technology is the signalling system. Not only does this control and determine how trains run, it also links in to many of the other technologies used by Crossrail.

Signalling, like the actual rails trains run on, is nothing new. In the UK most lines still rely on a largely mechanical system which divides routes up into blocks with track-side traffic light signals informing a driver if the next block is clear. Over the years systems such as TPWS (Train Protection Warning System) have been overlaid on to this set up to automatically apply train brakes should a service proceed past a signal on to a block which has not yet cleared.

Slowly, very, very, slowly, signalling in the UK is being modernised to support the European Train Control System (ETCS), a solution developed on the continent and designed to standardise signalling systems and improve interoperability between networks in different countries. This may be less of an issue in Britain, with only the Channel Tunnel connecting us to our European neighbours, but there are still myriad benefits including efficiency, safety and compatibility with equipment from different suppliers.

ETCS is often described in terms of levels and, for the UK, level 2 is seen as the game-changer. Trackside traffic lights are no longer required – radio waves and balises fitted between the track identify a train's position and communicate 'movement authority' as appropriate. In due course all main line railways in Britain are destined for ETCS level 2 with its promise of greater operating efficiency and cost savings from requiring less supporting trackside infrastructure.

Underground systems require something more. To run a metro service with trains leaving stations every two to three minutes requires more sophistication than a fixed block signalling system. The solution here is to implement moving block – ensuring a specified headway between trains is maintained rather than focusing on vacating a specific section of track. ETCS level 3 supports such a moving block system.

187

If building a new underground, metro railway ETCS level 3 appears the modern, European standard option. A line-wide ETCS could be specified across the route – level 3 perhaps could be specified in the core with level 2 on route sections with lower train frequencies.

In a world where consumer-focused technology develops at pace this assumption seems reasonable. Yet ETCS development has stuttered, complicated by sub-levels and add-ons designed to ensure safety, overcome incompatibilities and respond to specific line idiosyncrasies. Off-the-shelf, work-out-of-the-box solutions remain elusive with infrastructure operators reliant on specialist suppliers to adapt systems for particular routes. That introduces 'unknowns' into schedules and inflates the cost of commissioning.

ETCS, despite being an accepted way forward, has had a slow start in the UK. The first commissioning was the Cambrian line in Wales in 2011, a project fraught with challenges despite the low service frequencies on the route. Of more relevance to us is the implementation of ETCS level 2 on the Thameslink core through central London as part of the Thameslink Programme in 2018.[4] There have been plenty of plans to commission ETCS on main line routes, including the Great Western, but projected delivery dates have come and passed. At the time of writing work was underway to commission ETCS on the southern section of the East Coast Main Line.

This is a contemporary discussion but, remember, for the Elizabeth line the signalling specification was being decided in the early 2000s. The development team was well aware of ETCS, anxious to support mandated European interoperability regulations as well as building a railway that made use of the best, up to date technology. Yet, when looking at how trains could run through central London every few minutes, there was not much evidence of tried and tested ETCS systems to draw on. Level 3 systems were more conceptual than actual. Even ignoring the requirements of the Crossrail central

---

[4] Railway Gazette International, Thameslink first with ATO over ETCS, 20 March 2018

operating section there were few examples to give confidence that specifying ETCS was a sound choice.

It is to the credit of the development team that it decided, early on, that Crossrail would not be a 'bleeding edge' railway. When making plans it's easy to get starry eyed about technology – the Railtrack-era proposal for moving block signalling on the West Coast Main Line which has yet to be delivered springs to mind – but no, Crossrail said it would specify tried and tested technology. That position led to the Elizabeth line signalling system we have today.

## CBTC at the centre

At the heart of the new railway, the Crossrail central operating section, 42 kilometres of mainly underground track, is a communications-based train control system (CBTC) similar to those used on metro systems around the world including the London Underground Victoria, Jubilee and Northern lines as well as the Docklands Light Railway. Crucially, this provides a tried and tested, moving block signalling system to support high frequency train services. The central signalling system, including the short south-eastern overground stretch to Abbey Wood, is overseen by the Elizabeth line Route Control Centre in Romford. Drivers decide when to close the train doors and check platforms but computers handle the rest.

The CBTC signalling and control system, Trainguard MT, was supplied by Siemens and designed, manufactured, installed, tested and commissioned at Siemens Mobility facilities across the world. Initially tested in Braunschweig, Germany the system was then integrated with other elements of the signalling and the platform screen door systems at Siemens Mobility's manufacturing and testing facility in Chippenham, Wiltshire. Siemens was sub-contracted by train manufacturer Bombardier (later taken over by Alstom) to ensure systems fitted alongside the track communicated reliably with equipment in train cabs.

CBTC, the signalling system which ensures trains maintain a safe distance from each other, implements Automatic Train Operation.

ATO controls the train's traction system and brakes so as to automate starting, accelerating, cruising, coasting, braking and stopping. Smooth brake curves play a key role in this computer-based system; by comparing up to date track and timetable data – including where the preceding and following services are – with the train's acceleration and braking capabilities the system can calculate the optimum speed profile at any given time.[5] This means that the computer can drive the train better than any human driver to smooth the flow of multiple train services and maximise energy efficiency.

Elizabeth line trains run in ATO mode through the central operating section from Portobello Junction to Pudding Mill Lane/Abbey Wood. Although the initial service sees up to 24 trains per hour run in each direction between Whitechapel and Paddington during peak hours, the new signalling system is capable of enhancement to 30tph through the central section at a later date.

With its proprietary CBTC system, Elizabeth line core signalling in some ways looks old-fashioned compared to the Thameslink core with its ETCS plus ATO overlay. However, it's worth remembering that this is ETCS level 2 (not level 3 moving block), so would be unlikely to support the present 24 train per hour peak let alone the potential future 30. Also, when it went live in 2018, Siemens described the Thameslink core as the world's first operational application of ATO over ETCS.[6] How could the Crossrail planners, in the early 2000s, specify a technology that had yet to be deployed commercially?

## Overground alternatives

Beyond the new railway Elizabeth line trains run on Network Rail infrastructure, used by a variety of rolling stock which has yet to

---

[5] Siemens European Train Control System (ETCS), https://www.mobility. siemens.com/global/en/portfolio/rail/automation/automatic-train-control/ european-train-control-system.html
[6] Siemens European Train Control System (ETCS), https://www.mobility. siemens.com/global/en/portfolio/rail/automation/automatic-train-control/ european-train-control-system.html

deploy ETCS. On much of the Paddington to Reading and Liverpool Street to Shenfield sections trains use TPWS, an enhancement of the Automatic Warning System (AWS) which dates back to the 1950s. For these parts of the Elizabeth line drivers are in full control of their train unless they fail to slow down when required or overshoot a signal set to red. In such cases TPWS automatically applies the brakes. Signalling for Crossrail's western route is controlled from the Thames Valley Signalling Centre at Didcot and for the north eastern branch from Liverpool Street Integrated Electronic Control Centre (IECC); this will eventually transfer to Romford as the infrastructure organisation proceeds with plans to consolidate signalling nationwide.

To switch between signalling systems requires a sophisticated auto-transition system. Heading into the Crossrail central operating section drivers are asked if they would like CBTC to take control of the train and must press a button to accept. Heading out of the central section drivers are asked if they would like to take back control of the train and, again, must press a button to indicate that they accept this responsibility. In the west the transition occurs as trains travel through Westbourne Park. In the east acceptance of the transition happens while a train has stopped at Stratford station.

A further disparity emerged on the route from Airport Junction on the Great Western route to Heathrow Airport, through the Heathrow rail tunnels, built separately from Crossrail in the 1990s.[7] Older rolling stock, such as the Heathrow Connect Class 360s, relied on the Automatic Train Protection (ATP) system for signalling through the tunnels. Would the new Class 345 trains have to be fitted with ATP kit as well as CBTC and TPWS to reach Heathrow? This option was rejected after considering difficulties obtaining equipment from suppliers, ETCS compatibility concerns and complications associated

---

[7] Heathrow Express Tunnels, NEC Contracts (part of the commercial arm of the Institution of Civil Engineers), https://www.neccontract.com/projects/heathrow-express-tunnels-london-uk

with cab fitment.[8] Instead, this part of the Elizabeth line became one of the first lines in the UK to run using ETCS level 2 – but only after systems in the tunnel had been installed and train compatibility issues resolved.

Following the opening of the Elizabeth line Network Rail commissioned a new stretch of ETCS route from Airport Junction to Acton. Since going live on 27 November 2023 trains to and from Heathrow are now signalled using ETCS between Acton and the airport. Trains to Reading transfer from TPWS to ETCS at Acton before switching back to TPWS just after Airport Junction; the transitions also apply in reverse for trains heading into London. Network Rail hopes to eventually extend ETCS to include Paddington main line station but this presents considerable challenges.

To summarise, the Elizabeth line signalling system is in fact three separate systems: TPWS/AWS, CBTC for the central section, and ETCS for the Heathrow spur and between Airport Junction and Acton. ETCS is the 'master' overarching system and, back when the UK was part of the European Community, a derogation to allow the installation of CBTC (rather than ETCS) through the core was only permitted with the agreement that an upgrade to ETCS would be pursued once the technology became mature enough to support Crossrail services. When that happens, and when Network Rail gets round to commissioning ETCS on all overground lines from Paddington and Liverpool Street, the Elizabeth line may become a full ETCS railway. That appears unlikely to happen any time soon.

## Living with the decision

That's how we got here. Of course it would be nice, as with any new shiny railway, to have a fully integrated, route-wide signalling and train control system. For the reasons above the Elizabeth line does not achieve this. But while the solution chosen is not perfect, it is pragmatic.

---

[8] Provision of GW-ATP on Crossrail Class 345 trains, Crossrail Ltd, November 2014, https://www.orr.gov.uk/media/16408

Having three signalling systems has, undoubtedly, made it more difficult to develop a reliable train control system for Crossrail. Trains need to communicate with the new technology installed on the central operating section as well as Network Rail trackside kit, some of which has been in place for many years. Even when buying new equipment Crossrail has been unable to buy an 'off the peg' system and has had to pay programmers around the world to write the code that allows different systems to work together.

When one recalls the challenges faced during commissioning of the Elizabeth line (see chapter 9), or looks back through the papers of Transport for London meetings, mention after mention can be found of the latest software 'drop'; of plans to install the newest iteration of a program to resolve some shortcoming with the signalling or train control system. As one set of flaws was fixed more emerged. Eventually the software was mature enough to be used for passenger service but you can be sure there will be more incremental upgrades to come.

All this underlines an emerging truth: hardware is declining in importance. With a computer-controlled railway the power rests with software that governs how electronic components interact. That in turn hands power to the programmers, the specialists we rely on to translate our wishes into computer speak.

Maybe a unified, state-of-the-art signalling system would have been different; we will never know. However, all things considered, it is hard to imagine it would have been more straightforward.

For all the challenges faced, Crossrail signalling represents a remarkable achievement. The opening of the Elizabeth line marked the first time national rail lines to the east and west of London had been connected through an underground metro system. The transitions – where trains automatically switch between signalling systems – are crucial to running a safe and reliable service.

Incline lifts can be found at Liverpool Street (pictured)
and at the eastern end of Farringdon platforms.

# 9. THE CRISIS

Big rail infrastructure projects are notorious for missing delivery dates and coming in millions, sometimes billions, over budget. Scanning back over the major rail schemes delivered in Britain over recent decades it's not difficult to find evidence of this. The ultimately successful Thameslink Programme, originally known as Thameslink 2000, was repeatedly remapped until completion in 2019.[1] The West Coast Route Modernisation, which saw inter-city services to and from Euston transformed with a new fleet of tilting Pendolino trains, has yet to deliver the moving-block signalling promised in the early days of rail privatisation. And then there's High Speed 2.

There are many reasons for these cost rises: inflation, premiums being demanded for specialist skills and labour, inadequately specified projects, new and unproven technology, political pressures – the list goes on. Time and time again these and other facets intertwine to create rail projects that are complex and unwieldy.

This helps explain the reticence from government in committing funds to Crossrail[2] which would inevitably represent a draw on the Treasury over a prolonged period. Not only would large sums of public money be required but the amounts needed could increase over time. Those concerned that once construction began in earnest it might prove difficult to refuse extra funding were absolutely right.

So, from early on, Crossrail observers – myself included – were rightly watching the clock and the money. Was the programme on time and on budget? The answer from the Crossrail press office, and indeed from senior figures in Crossrail Ltd and Transport for London, was for many years a clear and definitive 'yes'.

---

[1] NR hails fruits of investment in final CP5 books, Transport Briefing, 26 July 2019
[2] See chapter 1

While this may well have been true early on in the construction phase we now know that the scheme was not delivered on time – that is the central section – the new-build Elizabeth line – was not ready to open in December 2018 as had been the plan since 2010.[3] This in turn meant the full completed cross-capital railway could not be achieved in December 2019. Neither was the infrastructure delivered on budget; by which I mean within the £14.8 billion funding envelope agreed by the government comprehensive spending review of 2010.

But for years Crossrail appeared to be doing well. Tunnelling contractors dug the vast tube tunnels for the new railway. Procurement proceeded at pace with contractors lined up to build the new stations. The National Audit Office's assessment of the programme in January 2014 praised the oversight of the project: "On the whole and to date, the Department [for Transport] together with its co-sponsor Transport for London and its delivery body, Crossrail Ltd, have done well to protect taxpayers' interests in the Crossrail programme. In the early years, they took effective action to stop costs escalating and to obtain more competitive rates from suppliers during the recession. During the construction phase, the governance arrangements and oversight of the project have ensured tight management of the programme so that delivery to both cost and schedule are well managed."[4]

So where did things start going wrong? Perhaps one of the first signs was that the contingency budget was being eaten into. In the March 2015 issue of Modern Railways, with tunnelling under central London at an advanced stage, Transport Commissioner Peter Hendy said that there was a 20% chance that some of the £600 million TfL contingency funding might be needed. According to the commissioner: "Crossrail's cost performance is being challenged by cost growth in a number of contracts. However, plans have been

---

[3] CSR: Crossrail survives but tunnelling reprogrammed, Transport Briefing, 20 October 2010
[4] First National Audit Office Crossrail report: Crossrail, 24 January 2014

made to address this as the major tunnelling and civil [engineering] contracts approach completion."

Fast forward a year to the March 2016 issue and the official line was that the project remained on time and on budget – within the £14.8 billion envelope agreed for the core infrastructure. Mike Brown, Hendy's successor as commissioner, said there was around a 30% chance that some of the £600 million TfL contingency funding might be required. We were told cost pressures that had emerged over the past year on an unspecified number of contracts continued to be addressed by Crossrail Ltd's senior management.

By 2017, with the programme officially more than 80% complete, confidence in progress was so high that work began to wind down Crossrail Ltd. Procurement and internal audit staff were transferred back into parent organisation Transport for London. Responsibility for operation of the Tunnelling and Underground Construction Academy (TUCA) at Ilford, where more than 16,000 Crossrail workers received training, transferred from Crossrail to TfL.[5]

A Crossrail Transition and Integration Programme was established and saw TfL identify Crossrail Ltd staff about to finish work on Crossrail and assess their interest and suitability for work within TfL on other major projects such as Crossrail 2. Day-to-day Crossrail Ltd activities were reorganised so that services such as IT support could be scaled down and transferred into the wider TfL structure as soon as this became more practical and cost effective than retaining a standalone operation.

At the time this seemed like prudent planning for the end game. The Crossrail delivery engine was vast and expensive engineering resources would surely be called upon less and less as the new railway was delivered.

---

[5] The £13 million academy opened in 2011 in response to concerns that there were not enough people with the necessary skills and experience to build Crossrail

But hang on, impressive as all those construction milestones seen in chapter 4 were, at this stage no railway had been delivered. Crossrail remained a notional construct – as a functioning railway it did not exist.

This was reinforced as the first of the five stages in the opening schedule approached. A Crossrail update to the TfL Board in February 2017 began by reasserting that "Crossrail remains on schedule to open, as planned, in five stages".

Those long-planned five stages were:

(1)    Liverpool Street high level to Shenfield in May 2017;
(2)    Paddington high level to Heathrow Terminal 4 (replacing Heathrow Connect) in May 2018;
(3)    Opening of Elizabeth line tunnels and stations through central London and Docklands (Paddington low level to Abbey Wood) in December 2018;
(4)    Joining the Shenfield branch to the through-London service in May 2019;
(5)    Joining the Heathrow and Reading branches to the through-London service in December 2019.

With the May 2017 stage one launch looming the board was told that, in order for the first train to be successfully introduced into passenger service, work still needed to be completed at Ilford depot as well as by Network Rail to adjust platform edges and provide equipment at stations on the Shenfield branch. Weekend closures were scheduled to allow this to be completed in March.

Of greater concern was the report of work still required to have stage two operational in May 2018 with four trains per hour between Paddington and Heathrow. For stage two services the new Class 345 trains needed to operate using the European Train Control System (ETCS) between Heathrow and Airport Junction.[6]

---

[6] See chapter 8

Network Rail was responsible for fitting trackside systems with Bombardier, as manufacturer of the new trains, fitting the required on-board equipment; this needed to be integrated and tested in the second half of 2017. Between Airport Junction and Paddington trains would use existing signalling systems.

21 May 2017, the date of the national rail summer timetable change, was when Crossrail train operator MTR was due to start running Class 345 trains between Liverpool Street high level and Shenfield[7] in accordance with the timetable laid out above. However, with testing of the fleet still underway no trains were ready. Howard Smith, TfL Operations Director for the Elizabeth line, said: "The train is still undergoing thorough testing, assurance and approvals before it enters passenger service shortly." Difficulties with platform equipment and train doors were subsequently cited as causes of the delay bringing the trains into public use.

Although Class 345s under test had become a familiar site on the Shenfield route it was Thursday 22 June when the first 345 was brought into passenger service. As testing continued the new trains gradually replaced British Rail-built Class 315s with 11 sets running on the route by the autumn.

Stage one launch was an inauspicious start. While commissioning trains takes time and patience this was arguably the easiest to achieve of Crossrail's five delivery milestones. As the year rolled on other pressures on the project emerged: Trade union Unite announced that electricians, mechanical engineers and plumbers working on Whitechapel station would receive productivity payments every four weeks. It was becoming clear that the trades had considerable bargaining power at this stage of the project – a situation that had echoes of the Jubilee line extension project when pressure on bosses to get the line open on time led to inflated wage demands.

---

[7] MTR took over the Liverpool Street to Shenfield stopping service previously run by Abellio Greater Anglia on 31 May 2015. See chapter 14 for further information about MTR

In July 2017, responding to a Freedom of Information request by the *Sunday Times*, Crossrail Ltd released a breakdown of the final cost of 43 (of 65) tier one Crossrail contracts.[8] This revealed remarkable increases in costs. Carillion's Paddington Integrated Project job (C272) was awarded for £28 million but the final cost was £98 million. Liverpool Street advance works (C501 and C503) more than doubled. The western tunnels contract (C300), awarded to BAM Nuttall/Ferrovial/Kier went from £473 million to £745 million and the eastern tunnels package (C305), contracted to Dragados/ Sisk, jumped from £479 million to £756 million.

While the awarded contract prices excluded employer risk, inflation and subsequent changes to scope (and these were substantial for the Paddington Integrated Project) it seemed fair to wonder whether the final value of the contracts – including the station and fit-out works which at the time were still underway – could be accommodated within the £14.8 billion Crossrail funding envelope. Crossrail told me: "With the project now 85% complete Crossrail Ltd has a very high level of confidence and visibility about the programme's final cost. Crossrail Ltd continues to forecast that the Crossrail programme will be delivered within its £14.8 billion funding envelope."[9]

At the Transport for London board meeting in November Commissioner Mike Brown noted that Crossrail Ltd was still forecasting that the programme would be delivered within its £14.8 billion overall funding envelope. He said the key risks continued to be management of cost and schedule pressures "on a small number of contracts, and Bombardier's software development and testing for the new Elizabeth line trains". This reference to the trains was crucial although at this stage its full significance was not clear.

---

[8] Crossrail confirms £1bn extra cost of major contracts, Construction Enquirer, 7 August 2017
[9] Two thirds of Crossrail final contract values published, Transport Briefing, 10 August 2017

Brown went on to flag a new area of concern – Network Rail. Although the schedule to have a fully operational Crossrail service running in December 2019 had been nailed down in 2010, by 2017 a significant tranche of upgrades planned for the overground Elizabeth line routes had yet to be procured. Brown told the meeting the main works tender to upgrade the western stations was issued in October 2017 to enable phased completion of the station upgrades between September-December 2019. This all looked distinctly last minute.

It was not just Network Rail delivery deadlines that were under pressure. Board members were also told: "Additional funding is required to ensure that the scope of the surface works is delivered in full, including station upgrades to the east and west."

The theme continued at TfL's Programmes and Investment Committee on 12 December. TfL said the development and assurance of the train signalling and software remained the most significant single risk to stage two services starting on time because successful completion of testing had to be followed swiftly by submission of assurance information to the relevant authorities and the training of more than 60 drivers. Yet testing of ETCS kit, which was being fitted to the Class 345s, was behind plan due to what TfL described as the "immaturity and instability of train software".

Meanwhile, the energisation[10] and opening of the new Old Oak Common depot was moved to the start of January 2018 rather than late November 2017 as previously planned. In November a loco-hauled Class 345 became the first Elizabeth line train to visit the depot.

## Optimism checked

2018 is when the positivity that had so far accompanied the Crossrail enterprise vanished. After delays to the stage one launch it was becoming clear that stage two, scheduled for May 2018, was not going to go as planned. The intention had been to replace the

---

[10] Power to overhead wires above the train tracks into and outside the depot being switched on

Heathrow Connect stopping service and double the frequency of local services between Paddington main line station and Heathrow Terminal 4 from two to four trains per hour. Class 345s, operated by MTR Crossrail under contract to TfL, were to replace Class 360 Connects.

Well before May it was clear the 345s were not ready – specifically the ETCS equipment and software installed on the trains which was needed to operate through the Heathrow tunnels to reach the airport. To mitigate the delay MTR Crossrail retained Class 360s, when it took over the route on Sunday 20 May, to provide a stopping service to the airport but with the current two trains per hour frequency. Two additional trains an hour were operated using Class 345s but terminated at Hayes & Harlington station given that ETCS was not yet reliable enough to guarantee a regular service onwards to the airport. To compound the difficulties it looked as if the new bay platform at Hayes & Harlington would not be commissioned in time by Network Rail and consequently short formation 7-car Class 345s were initially used to be replaced by 9-car sets once the building work was done.

In an update to the TfL Board on 30 January regarding Elizabeth line operational readiness Howard Smith, now TfL's Crossrail Operations Director, noted that the development and assurance of the train signalling and software remained the most significant single risk to timely commencement of stage two services. As described above stage two was now set to go ahead with a hybrid 360/345 service and Smith told the meeting that the full stage two service would be in place by autumn 2018. The stage two launch thus moved from the envisaged step change to the start of a gradual commissioning of Class 345s west of Paddington.

If a little bit of ETCS in the Heathrow tunnels was proving sticky for stage two, how would Crossrail cope with the stage three launch six months later when Class 345s were supposed to be available for use in the new and much longer Elizabeth line tunnels with another new signalling system – this time Communications Based Train Control – software for which remained under test? While the trains did not need ETCS and CBTC working at the same time until stage five, testing of CBTC alongside other complex electronic systems had yet to begin.

It had been clear early on that ahead of the stage three launch a generous period for train testing in the tunnels would be required. But a major problem with a new high voltage transformer at Pudding Mill Lane would have a significant impact on the time available. According to Crossrail Ltd: "During the initial energisation of electrical equipment at Pudding Mill Lane sub-station on 11 November 2017, two voltage transformers failed." Crossrail Chairman Terry Morgan told the TfL Board: "It got switched on – and exploded."

Successive attempts to turn on the power also failed and put back train testing by at least three months. Howard Smith wanted a minimum of three months of trial running but even that would have to fit around other workstreams; while trains were running through stations the work that could happen at platform level was limited. By 2018, looking at the opening schedule, you might imagine station works would be close to completion. They were not.

The pressure on Crossrail was building with the significant challenges facing the programme now being voiced in public. While soft, phased launches for stages one and two had been possible that would be much more difficult for stage three. Still, Morgan told the board he remained "very confident" that the launch of the new railway would go ahead in December. On the issue of money the Crossrail chair informed the board "we are very close to the funding envelope". Mark Wild, London Underground Managing Director, added: "We can still do it, but it's very, very hard and complex and it brings with it cost pressures as well."

The electrical sub-station at Pudding Mill Lane was successfully switched on – and with it the first 25KV overhead lines – on 1 February 2018. This meant that by the end of the month Crossrail was able to announce that an Elizabeth line train had travelled through one of the new Crossrail tunnels for the first time without a locomotive.

With this issue resolved test running could get going. Hundreds of runs would be needed to rack up the miles and assurances required to allow Elizabeth line trains to enter passenger service and ensure the CBTC system would work without interfering with other electrical systems.

Meanwhile, Bombardier launched operations at the new Elizabeth line depot at Old Oak Common. The skeleton staff at the depot was expected to increase to around 80 people by the summer as Bombardier delivered more 345s and the train testing programme gathered pace.

Ahead of the planned stage three launch Crossrail Ltd Chief Executive Andrew Wolstenholme left the Crossrail project after seven years to take up a new role in the private sector.[11] His departure left Crossrail Programme Director Simon Wright in charge in a new combined role as Chief Executive and Programme Director. Wolstenholme would continue to work with the Department for Transport in his capacity as a non-executive Director of HS2 Ltd.

In July, on the final day before the Westminster parliamentary summer recess began, transport minister Jo Johnson – brother of the more famous Boris – issued a written statement providing an update on Crossrail finances (a mandatory annual update was stipulated as part of the Crossrail Act). The 2018 update revealed that 2017's reported forecast by the Crossrail Board that the cost of constructing Crossrail would be within the overall £14.8 billion funding envelope – the scheme was 'to cost no more than £14.5 billion' – was incorrect. Johnson said that, following agreement between the Department for Transport and Transport for London, that figure – settled upon in 2010 – had been increased by £590 million to give a new funding envelope of £15.4 billion.

---

[11] Wolstenholme to relinquish top job at Crossrail, Transport Briefing, 8 March 2018. Andrew Wolstenholme worked for Balfour Beatty before being appointed Crossrail Chief Executive in 2011.

Crossrail Ltd said £290 million (from the DfT/Network Rail) of the £590 million would go to Network Rail to complete its 'on-network' upgrades to the existing railway. The need for extra money was attributed to the complexity of integrating new technology on the national rail network, such as a new driver only operation camera system, and the poor underlying condition of Victorian infrastructure.

Of the remaining £300 million heading to Crossrail Ltd, this would be channelled to systems integration and testing. Crossrail Ltd said increased time and resource was required for the installation and testing of the multiple systems needed to operate the new railway, including track, communications, power and signalling, and worse than expected ground conditions encountered during station construction. These were said to have prolonged works and compressed the delivery schedule. It added that some construction costs were higher than originally forecast.

## Reality revealed

On 31 August 2018 the announcement finally came: stage three opening – creating the cross-London Elizabeth line – would not be achieved as planned on 9 December 2018. The announcement followed an extraordinary meeting of the Crossrail Ltd Board on 29 August when members agreed that it was no longer possible to deliver a safe and reliable railway for passengers in time for the planned 9 December opening. A partial opening was considered but deemed impractical.

Press reports, TfL meetings (both public and private sessions) plus industry rumours meant this was hardly a surprise. Yet to hear, less than four months before launch date, an official acknowledgement on the last day of the parliamentary recess that the opening date planned for since 2010 would be missed – despite repeated assurances from Crossrail Ltd that all was well – was quite a moment.

TfL explained: "The original programme for testing has been compressed by more time being needed by contractors to complete fit-out activity in the central tunnels and the development of railway systems software. Testing has started but further time is required to complete the full range of integrated tests."

Underestimating the systems integration work required for the project has since emerged as a major cause of the delay to the Elizabeth line opening. Bombardier and signalling contractor Siemens were struggling to get train and tunnel systems to communicate effectively. Software that allows the Aventra trains to switch between CBTC in the central section and TPWS west of Paddington and east of Liverpool Street was described as 'immature' with extensive development work still required.

But the hold-up was not all about tech; across the project there was plenty of building work that simply was not finished. Even if the trains and signalling had been ready Bond Street station was nowhere near ready for a December 2018 opening.

Different delays fed into each other. The explosion of the transformer at Pudding Mill Lane in October 2017, halting train testing and prompting safety investigations, was clearly not helpful. Yet this incident was a contributing factor rather than the sole reason for the opening delay. When tests of the Class 345 trains resumed these were hampered by incomplete station infrastructure. This limited test runs and suggests that even without the transformer fault Crossrail would have struggled with its train commissioning programme.

Following the exit of Crossrail Chief Executive Andrew Wolstenholme in April it fell to his successor, Simon Wright, to make excuses and reflect on the progress that had been made. As autumn arrived what had, only weeks before, been presented as the final activity to deliver Crossrail reemerged as the beginning of a new phase in the programme with no predetermined length. We were told: "TfL and Government continue to work through the financial and other implications of the delay in Crossrail Central Operating Section opening."[12]

---

[12] TfL Programmes and Investment Committee meeting, 11 October 2018, https://tfl.gov.uk/cdn/static/cms/documents/pic-20181011-agenda-and-papers.pdf

A month and a half before stage three of Crossrail had been due to open came news that Mark Wild, London Underground's Managing Director, had been appointed as the new Chief Executive of Crossrail Ltd, the third in 12 months. Commenting on the appointment of his successor Simon Wright said: "Everyone involved in the project is fully focused on ensuring the Elizabeth line is completed as quickly as possible. With the construction phase due to come to an end later this year, the major focus for Crossrail will be the integration of the complex railway systems and the start of full-time testing."

This statement underlines how behind schedule the programme was. A month and a half before stage three of the Elizabeth line had been due to open construction work had yet to finish and full-time testing had not begun.

A Heathrow Connect Class 360 passes the new Hammersmith & City line ticket hall being built by Carillion as part of the Paddington Integrated Project.

# 10. THE EXTRA MONEY

The funding components of Crossrail are set out in detail in chapter 2. This explains how, as a result of the government's comprehensive spending review in October 2010, the programme came to have a 'funding envelope' of £14.8 billion. This budget survived until July 2018 by which time, as we saw in the previous chapter, it was clear that statements claiming the Elizabeth line was 'on time, on budget' were not true.

In May 2019 the National Audit Office reported that the total funding envelope for Crossrail had increased to £17.6 billion.[1] This comprised two increases in funding, released as the opening schedule and budget for the Elizabeth line was revealed in 2018 to be inaccurate. £590 million was agreed in July 2018, including £290 million for Network Rail, and a further £2.15 billion in December 2018.[2]

In November 2019 Crossrail Ltd parent Transport for London advised the stock exchange that it would need another £400 million. But by August 2020, after Covid-19 restrictions had delayed the project further, the Crossrail board reported that it could need up to £1.1 billion to complete work on the scheme. On 30 November 2020, the Department for Transport agreed that the Greater London Authority would be allowed to lend TfL an additional £825 million to allow completion of the Elizabeth line. The loan was funded by borrowing against future London Community Infrastructure Levy and business rate receipts which had exceeded early Crossrail projections but had previously been viewed as a source of funding for Crossrail 2. The £825 million was split between a £500 million loan with a fixed repayment profile and a £325 million loan to be

---

[1] Second National Audit Office Crossrail report: Completing Crossrail, 3 May 2019
[2] A detailed breakdown of funding sources to April 2019 is available in Figure 6 of the National Audit Office report, A memorandum on the Crossrail programme HC 1924

repaid subject to CIL and business rate supplement revenues. London's Transport Commissioner said it might take until 2043 for the loan to be paid back.[3]

Not only were the costs overseen by Crossrail Ltd rising but Network Rail's parts of the build were also exceeding budgets. Further increases to the forecast cost of completing works on the national rail network meant that, on top of the extra £290 million authorised in July 2018 (see above), Network Rail provided additional funding in July 2019 and July 2020 totalling £390 million. This came at the expense of railway spending elsewhere: money was repurposed from underspends and efficiencies from Control Period 5 (£250 million) and a change to Network Rail's spending plans for Control Period 6 (£140 million).[4]

The cost of Network Rail's Crossrail work increased for multiple reasons including changes in scope and because the assets that it was working on were in a worse condition than anticipated. A contractual dispute with one of its main contractors didn't help[5] and Network Rail had to replace Carillion as a contractor when the company went into liquidation in January 2018. Costs also increased because of delays resulting from the re-tendering of several significant contracts and because additional work had to be carried out. Even work previously planned took longer to complete than expected, including upgrades to power systems on the eastern section.[6]

[3] House of Commons Committee of Public Accounts Crossrail: A progress update Twenty-Fourth Report of Session 2021–22
[4] Railway five year funding terms. CP5 ran from April 2014 to the end of March 2019 with CP6 from April 2019 to March 2024
[5] Second National Audit Office Crossrail report: Completing Crossrail, 3 May 2019
[6] Third National Audit Office Crossrail report: Crossrail – a progress update, 9 July 2021

The result of these changes was that by May 2021 the total funding envelope, the money available for the programme, was £18.8 billion. The updated funding breakdown, setting out where the different components of the funding package came from, is set out below.

### Crossrail funding breakdown at May 2021

| Source | Amount (£ million) |
|---|---|
| Department for Transport | 5,110 |
| Business Rate Supplement, borrowing and direct London contribution | 4,100 |
| Network Rail | 2,980 |
| DfT loan to GLA funded from Business Rate Supplement and Mayoral Community Infrastructure Levy | 2,125 |
| Transport for London | 2,050 |
| DfT loan to TfL funded from future Crossrail revenue | 750 |
| Sale of surplus land and property | 550 |
| Community Infrastructure Levy | 300 |
| Developer contributions | 300 |
| City of London Corporation (committed) | 250 |
| GLA | 100 |
| London businesses (voluntary) | 100 |
| Heathrow Airport | 70 |
| TOTAL | 18,800 |

*Source: Transport for London*

When the House of Commons Public Accounts Committee published *Crossrail: A progress update* in October 2021[7] Crossrail blamed "schedule delay" for the bulk of the cost rises, although it said the

---

[7] House of Commons Committee of Public Accounts Crossrail: A progress update, Twenty-Fourth Report of Session 2021–22

figure also included £228 million of Covid-19 related spend. The Department for Transport attributed the £390 million Network Rail increase to historical issues, such as adding previously missing scope, lower than expected productivity of contractors, and impacts of the Carillion liquidation such as the time and money required to re-tender contracts.

Crossrail Ltd had relatively little exposure to the Carillion problems; although the company was main contractor for the Paddington Integrated Project (C272) to relocate the Paddington taxi rank to the canal side of the Brunel-built terminus (to enable construction of the new Crossrail station under Departures Road) this was completed well before the liquidation. However, while Network Rail awarded Carillion contracts for track, station and depot work on the Crossrail route west of Paddington years before the company ran into trouble, delays to site start dates meant Carillion had yet to fulfil its Crossrail commitments when the money ran out.

## Final account

In his 2022-23 budget Mayor of London Sadiq Khan agreed to provide an extra £48.5 million of GLA funding to complete Crossrail. The Mayor's contribution was matched by further additional funding from the DfT of £50 million, agreed as part of the August 2022 TfL funding settlement,[8] taking the Crossrail funding envelope to £18.9 billion. By this point, shortly after the opening of the Elizabeth line in May 2022, approximately £925 million of the 'up to £1.1 billion' needed had been identified.

The final stage of work to deliver the full Elizabeth line was characterised by modest fluctuations in cost forecasts. Yes, the full £1.1 billion extra TfL had said in 2020 that it needed had not been fully identified but, as we have seen with nearly £925 million secured, the gap had narrowed significantly. TfL was confident that savings

[8] Mayor of London's draft 2023-24 budget

211

and other financial workarounds would allow the programme to be completed even if the funding gap had not been entirely plugged.

At TfL's final Elizabeth line committee meeting, in July 2023, acting chief financial officer Patrick Doig stated that the 'anticipated final Crossrail direct cost' for the central section was £15.903 billion with the total funding package for the programme amounting to £15.8875 billion, leaving a funding gap of £15.5 million.[9]

### Increases in Crossrail funding since January 2014

| Figures in £m | Jan 2014 | July 2018 | Dec 2018 | May 2019 | Dec 2020 | May 2021 |
|---|---|---|---|---|---|---|
| Funding for Crossrail Ltd | 12,480 | | | 14,960 | | 15,790 |
| Increase | | 300 | 2,150 | | 825 | |
| Funding for Network Rail | 2,300 | | | 2,590 | | 2,980 |
| Increase | | 290 | | | 390 | |
| Total funding for Crossrail programme | 14,780 | | | 17,570 | | 18,770 |
| Expected opening date of central operating section | Dec 2018 | Autumn 2019 | No commitment made | Oct 2020 to March 2021 | Dec 2021 to June 2022 | Jan 2022 to June 2022 |

*Source: National Audit Office[10]*

---

[9] TfL Elizabeth line committee meeting, 25 July 2023
[10] Third National Audit Office Crossrail report: Crossrail – a progress update, 9 July 2021

## Timeline of total Crossrail funding allocation

| Date | Funding (£ billion) |
| --- | --- |
| 2007 | £15.9 |
| 2009 | £17.8 |
| 2010 | £14.8 (£12.5 Crossrail Ltd, £2.3 Network Rail) |
| July 2018 | £15.4 (£12.8 Crossrail Ltd, £2.6 Network Rail) |
| December 2018 | £17.6 (£15 Crossrail Ltd, £2.6 Network Rail) |
| July 2019 | £17.8 (£15 Crossrail Ltd, £2.8 Network Rail) |
| July 2020 | £18 (£15 Crossrail Ltd, £3 Network Rail) |
| November 2020 | £18.8 (£15.8 Crosrail Ltd, £3 Network Rail) |
| August 2022 | £18.9 (£15.9 Crossrail Ltd, £3 Network Rail) |

According to the National Audit Office, it may prove difficult to track the final cost of completing the Elizabeth line. Ending the main station work contracts was seen as key to reducing expenditure and as part of a staged handover Crossrail Ltd transferred some outstanding work to Rail for London Infrastructure and London Underground. It was not clear whether this work was to be completed at a later date or left if not considered critical to run the railway.[11]

To conclude this analysis of the Crossrail funding breakdown on a tidier note it's worth remembering that the Elizabeth line trains and Old Oak Common depot were paid for outside the figures quoted above and, as described in chapter 2, have a publicly quoted cost of circa £1.1 billion. Adding that on to the £18.9 billion August 2022 infrastructure figure in the table means it is not then unreasonable to say that it has, in total, cost around £20 billion to deliver the Elizabeth line.

## Political discord

To deliver a project as ambitious as Crossrail requires organisations to work together. It has to be said that this collegiate approach was

---

[11] Third National Audit Office Crossrail report: Crossrail – a progress update, 9 July 2021

notably absent during the later stages of the project when the sponsors were represented by Labour Mayor of London Sadiq Khan and Conservative Grant Shapps, Secretary of State for Transport from 4 July 2019 to 6 September 2022.

Khan, who presided over Transport for London, had seen his organisation's ticket revenue decimated by the Covid-19 pandemic, forcing him to seek a bailout from government. This was forthcoming, although only after the Mayor had threatened to wind down the capital's public transport services, but saw Khan forced to sign up to a raft of government conditions and policy initiatives.

To a certain extent Crossrail was protected; the project still had cash in the bank when TfL needed its first Covid bailout but not enough to complete the programme. And however much disdain there was for the Labour Mayor, allowing the project to lie unfinished was likely to reflect badly on government. So more money would be provided – but as little as Shapps could get away with.

When the £825 million loan was agreed at the end of 2020 Sadiq Khan said the government had insisted London must pay the Crossrail shortfall despite the majority of the tax income that would result from the new railway being destined for the Treasury. Shapps countered that London – as the primary beneficiary of the new railway – must ultimately bear any extra costs.

And so it seems that much of the overrun – the additional cost of delivering the Elizabeth line – will be borne by Londoners; recouped through fares used to pay back loans and paid in taxes by businesses and property developers; money that might have been available for a different purpose.

By the time the Elizabeth line opened in May 2022 there was a possessiveness about who had spent what: "This £9 billion investment in the Elizabeth line is the latest example of the government's commitment to London" proclaimed the Department for Transport.[12] The Transport for London press release accompanying the launch stated: "London is paying for most of the Elizabeth line, with nearly 70% of the total funding paid by London – made up of roughly 30% from London's farepayers, around 40% from London's businesses – combined with 30% from government.[13]

## Value for money?

To determine whether or not a transport project represents good value the Department for Transport has encouraged scheme promoters to present a benefit:cost ratio – essentially what the benefits of going ahead will be for every pound spent. In 2011 sponsors and Crossrail Ltd expected Crossrail to produce £1.97 of benefits for every pound spent on building, maintaining and operating the railway, or £3.10 including wider benefits from increased economic activity following completion. However, programme cost increases impact benefit calculations. In April 2019 the Sponsor Board examined indicative analysis of the potential effects of the cost increases and delays on the benefit:cost ratio for the programme. The analysis showed that Crossrail could produce around £1.50 of benefits for every pound spent, or around £2 including wider benefits.[14]

Benefit:cost ratios do not include the indirect economic impacts of a scheme such as building the Elizabeth line. Crossrail Ltd's Sustainability Summary 2018 said that the programme delivered more than 1,000 apprenticeships and 4,706 jobs for local/unemployed people, and that it has supported 55,000 jobs during construction with 62% of suppliers outside London.[15] There have also been

---

[12] Cheaper tickets and quicker journeys as part government funded Elizabeth line officially opens, DfT press release, 24 May 2022
[13] All aboard the transformational Elizabeth line, TfL press release, 24 May 2022
[14] Second National Audit Office Crossrail report: Completing Crossrail, 3 May 2019
[15] Third National Audit Office Crossrail report: Crossrail – a progress update, 9 July 2021

environmental benefits, such as the development of a new nature reserve at Wallasea Island using material excavated from tunnels.

## Learning lessons

Crossrail enjoyed a prolonged period as the poster child of rail infrastructure projects, bounding from one success to another. Ultimately, however, it's another scheme that – for all its impressive achievements – has come in late and over budget; from being an exemplar for several years it is exasperating seeing how it went wrong.

In situations such as this glib comments are often made about 'learning lessons' yet we must strive to do just that if there is to be any hope of avoiding a recurrence with future projects; indeed if there are to be future projects at all.

Reflecting on Crossrail finances over a 15 year period it's fascinating how the numbers were put together to create a viable project, the ongoing scrutiny, the cost pressures and the changes that had to follow. A vast amount of time was spent negotiating arbitrary figures which, it turns out, were purely hypothetical. It's one thing to specify savings but that doesn't mean they will be delivered.

When the Crossrail programme was conceived many people worried about runaway costs and it turned out they had a point. Once most of a new railway has been built a government cannot easily abandon a project – somehow it must be finished. Instead, when money ran short, the government – and Londoners - had to repeatedly shell out to find some more.

With the Elizabeth line now open it's easy to brush aside these difficulties, eclipsed as they are by all the achievements of the new railway. Unfortunately there has been an impact in reinforcing the views of those reluctant to commit to long-term infrastructure projects. It's easy now to forget that for a while a green light for Crossrail 2 seemed imminent, buoyed by the momentum achieved with Crossrail engineering successes. When I interviewed project boss Andrew Wolstenholme in 2013 he said: "If I was asked to do another 42 kilometres of tunnels we would not need to change the organisation too much.[16] Regrettably, delays delivering the Elizabeth line, and the need to draw on development funding that might once have been earmarked for Crossrail 2, mean the project is unlikely to be delivered before the 2040s.

---

[16] Modern Railways Crossrail supplement 2013

# 11. THE POST MORTEM

Building a new railway – underground – under London – was always going to be a huge undertaking, complicated and fraught with risk. Yet it was painstakingly planned, over many years, and drew on the experiences of building infrastructure schemes around the world. Overall, the programme of work was as expected.

So it is frustrating that the impressive end result has been marred by cost rises and delays. These are not minor, as might be reasonable to expect with a project of this size – the Elizabeth line was three and a half years late opening and at least £4 billion over budget.

A programme of works on this scale is vast and – let's be realistic – once begun has a momentum that is difficult to control. But while we can make excuses it would be much more useful if we could figure out where and why Crossrail ran into difficulties. If we can pinpoint what went wrong and learn lessons then future rail schemes have a much greater chance of going ahead.

## Contract issues

Crossrail's procurement strategy, explained in chapter 3, didn't work. Whether it was the best approach for the project can be debated but, with hindsight, we can see the target costs for contracts had little grounding in reality.

In the following analysis of what went wrong two issues repeatedly crop up. Scope of work changes; just as when fitting a new kitchen there's often something that wasn't foreseen so, in a similar way, alterations needed to be made to Crossrail contract terms. Secondly, most of these contracts had significant interfaces; contractors could not work in isolation – stations, tunnels, systems overlapped and progress on one contract was often dependent on progress with others.

These interfaces presented significant challenges, particularly as the programme schedule became squeezed, but ultimately it was more a problem for Crossrail than the contractors. The National Audit

Office found that changes to the design of construction and systems installation work, and changes to contractors' delivery schedules, cost around £2.5 billion between 2013 and 2018. "Crossrail Ltd did not require individual contractors to manage interfaces with other contractors, and so protected contractors from changes that were outside their control. Therefore, Crossrail Ltd had to compensate individual contractors for delays that occurred on other contracts, on which their work depended, and had to engage in costly change control negotiations.[1]

So while changes to scope added extra costs to contracts, the impact of the changes on other contracts exacerbated costs. The NAO noted that settlement of accumulated 'compensation events'[2] with contractors accounted for nearly £1 billion of the £2.5 billion cost increases. While chapter 3 noted limited disputes with contractors, at least during the early phase of Crossrail work, by January 2015, contractors had raised 16,000 notices with a backlog of 1,000 yet to be assessed. By January 2016 the number of notices had increased to 21,000 with 1,800 still to be decided. While some uncertainty due to commercial discussions is manageable and inevitable on major programmes, high levels can lead to time being absorbed managing commercial elements of the contract, rather than carrying out productive work. The build-up of compensation events on Crossrail was an indication of the high number of interfaces between contractors on the programme, and the prevalence of delays and change caused by poor integration of the work of multiple contractors.[3]

---

[1] Second National Audit Office Crossrail report: Completing Crossrail, 3 May 2019

[2] Under the terms of the target price contracts that Crossrail Ltd agreed with most of its main contractors, where cost increases occurred due to events outside a contractor's control they issued notification of a compensation event to Crossrail Ltd. Examples of such events included access to a work site being delayed, or a design change being instructed by Crossrail Ltd. The compensation event notice required Crossrail Ltd and the contractor to reach agreement about the nature of the change and any increase to the contractor's target price for the contract.

[3] Second National Audit Office Crossrail report: Completing Crossrail, 3 May 2019

Costs on most of the 36 main contracts listed in chapter 3 increased substantially. According to the NAO, between 2015 and 2019 there was little pressure on key contractors to deliver the programme efficiently.[4] During 2015 and 2016 some key contracts were moved from a target price to a cost reimbursement basis. This change meant that Crossrail Ltd removed the key incentive for contractors to minimise costs and took on the financial risk itself. The frequent re-planning of the programme, combined with increasing interfaces between contracts, meant that contractors continued to raise compensation events, and costs continued to increase. After it had announced that it would not open the central section in December 2018, Crossrail Ltd began negotiating fixed price contracts for some of the remaining work to improve certainty about costs. However, this risked removing commercial levers to ensure that contractors prioritised completion of Crossrail over other projects they were committed to.

The following table shows how many of the main works contracts saw significant cost rises. Some of the biggest increases were with the contracts to install track and key systems in the tunnels and some of the new stations such as Bond Street, Paddington and Whitechapel; the latter as a result of difficulties building around existing London underground and overground lines and station architecture. The extent of these complications appears not to have been recognised when the revised station design was revealed in 2010.

## Crossrail contract cost increases

When the National Audit Office published its third report on Crossrail in July 2021 the total cost of the 19 main works contracts still in place had increased by £1.3 billion. Six of the 36 main works contracts accounted for 74% of this forecast increase in main works costs between December 2018 and March 2021. Three stations accounted for over 40% of the total increase. The report also showed that it had been difficult for Crossrail Ltd to manage costs on the communication and control systems, and railway signalling contracts for the central section.

---

[4] Second National Audit Office Crossrail report: Completing Crossrail, 3 May 2019

# Crossrail tier one contract cost increases

| Contract | Number | Target at award (£m) | Forecast cost December 2018 (£m) | Forecast cost March 21 (£m) | Increase from award to December 2018 (%) | Increase from award to March 2021 (%) |
|---|---|---|---|---|---|---|
| *Tunnelling* | | | | | | |
| Eastern tunnels | C305 | 484 | 730 | | 51 | |
| Western tunnels | C300 and C410 | 490 | 749 | | 53 | |
| Thames tunnels | C310 | 196 | 229 | | 17 | |
| Station tunnels east | C510 | 246 | 510 | | 107 | |
| Pudding Mill Lane Portal | C248 | 52 | 184 | | 254 | |
| Eleanor Street and Mile End shafts | C360 | 46 | 255 | | 454 | |
| *Station main works* | | | | | | |
| Farringdon | C435 | 239 | 634 | | 165 | |
| Liverpool Street | C502 | 147 | 374 | | 154 | |
| Paddington | C405 | 181 | 571 | 649 | 215 | 259 |
| Bond Street | C412 | 126 | 412 | 660 | 227 | 424 |
| Whitechapel | C512 | 110 | 659 | 831 | 499 | 655 |
| Tottenham Court Road | C422 | 98 | 282 | | 188 | |
| Woolwich | C530 | 70 | 234 | | 234 | |
| *Systems* | | | | | | |
| Systemwide (track and electrical fit-out) | C610 | 323 | 956 | 1173 | 196 | 263 |
| Platform screen doors | C631 | 27 | 63 | | 133 | |
| Signalling | C620 | 51 | 131 | 236 | 157 | 363 |
| Communications and control | C660 | 43 | 139 | 263 | 223 | 512 |
| *Others* | | | | | | |
| Ilford stabling sidings | C828 | 54 | 153 | | 183 | |

Eastern and western tunnel contracts were completed prior to December 2018 and show their final values. Thames Tunnel and Pudding Mill Lane contracts show the costs when they were 98% and 99% completed respectively.

Target at award denotes the anticipated cost of the contract at award and includes adjustment for risk. Other cost values are contractors' forecast or final costs which may include adjustments for cost risks where the contract was not yet complete. Values drawn from Crossrail Ltd board reports. All values are in cash prices.

*Source: National Audit Office Crossrail reports 2019 and 2021*

Add in the £228 million impact of Covid on Crossrail[5] and the assessment of the project identified a total cost increase of £1.5 billion since April 2019. Sixty-two per cent of the total cost increase was attributed to work taking longer than expected.

## Crossrail management

Difficulties with contract interfaces were compounded by flawed planning of the Crossrail programme prior to Mark Wild's arrival and subsequent reorganisation (see chapter 12). Amazingly, it was not until late 2018 when Crossrail Ltd started to produce a detailed, realistic, plan.[6] Prior to this, from 2015, it had based its management of the programme on an aspirational plan designed to improve progress by suppliers, rather than to provide a reality check on overall progress. This plan did not adequately reflect interdependencies across the programme and therefore Crossrail Ltd did not have a full understanding of the risks to meeting the December 2018 Elizabeth line opening date.

Although the organisation took steps to address the mounting compensation events and renegotiated some of its key contracts to settle historical compensation claims, the complex interfaces between contracts remained and led to further delays. Delays to contracts also meant contractors' workforces needed to be paid for longer.

---

[5] See chapter 12
[6] Second National Audit Office Crossrail report: Completing Crossrail, 3 May 2019

The cost of the main Whitechapel station
construction contract increased more than 600%.

Overriding this was a confidence within Crossrail management that
the skills and experience of those leading project teams were so
exceptional they could deal with whatever challenges presented
themselves. According to Amyas Morse, head of the National Audit
Office: "Throughout delivery, and even as pressures mounted,
Crossrail Ltd clung to the unrealistic view that it could complete the
programme to the original timetable, which has had damaging
consequences... there have been a number of choices made in the
course of this project that have clearly damaged public value."[7] In
the unpicking of further cost rises contained within its third report
on Crossrail the NAO confirmed that the Elizabeth line was further
from completion than Crossrail Ltd realised when it set the revised
cost and schedule in April 2019. Indeed, the report raises significant
questions about the capability of the previous management: as well
as not understanding how near (or far) from completion the scheme

[7] Second National Audit Office Crossrail report: Completing Crossrail,
3 May 2019

was Crossrail Ltd did not understand the work required to bring a digital railway into service when it set its April 2019 plan.

Throughout 2019 and 2020 Crossrail Ltd repeatedly uncovered problems (that it had to resolve) with the assets already constructed. Previous management information did not provide an accurate picture of completeness. It was discovered that some assets needed more work for a range of reasons: the work had not been done; the physical asset was different from that documented; the work no longer met current regulations – for example, wiring in some stations and fire systems at Canary Wharf; or they were faulty – for example, fire doors.

With hindsight, decisions to allow bespoke designs of stations and limited standardisation across common assets, such as fire doors or CCTV, can be seen to have increased the complexity of integrating systems. However, other information that has emerged suggests the management of Crossrail pre-2019 had a poor grasp of the programme it was supposed to be managing. For example, management information used by Crossrail Ltd before 2019 did not include measures on completeness of documentation, key to ensuring assets could be brought into service, focusing instead on completeness of construction, making it difficult for the new Crossrail Ltd team to know accurately how complete the programme was when it set its April 2019 plan.[8]

Perhaps the clearest example of poor judgment was the premature decision to scale down the central programme and risk management teams during 2018 in anticipation of the programme reaching completion in December 2018. Why? This seems bonkers given the evidence from multiple sites that work was nowhere near finished. In 2019 the National Audit Office found Crossrail had 143 vacancies in its central delivery teams to work with main contractors to deliver the programme plus a further 33 vacancies

---

[8] Third National Audit Office Crossrail report: Crossrail – a progress update, 9 July 2021

across other areas.[9] Frustrating for those tasked with finishing the programme but not bad for those recently paid off specialists who found themselves once more in demand.

## Programme partners

As we saw in chapter 3 Crossrail brought in private sector expertise to rapidly create a fully-fledged construction unit. In March 2009 Transcend was appointed programme partner, responsible for strategic programme management, and a month later Bechtel was brought on board as project delivery partner, responsible for coordinating the activities of other contractors on the central section of what would become the Elizabeth line.

The 2009 project delivery partner contract required Bechtel to provide services against a scope, an annual service delivery plan and a schedule of deliverables. But in August 2019 Transport for London concluded that recent changes to the delivery schedule made the previous contract incentives ineffective.

A similar situation prompted TfL, in 2021, to approve a new incentive scheme for Transcend. The original programme partner contract had two components which determined how much Transcend got paid: (1) the time spent by staff it supplied to provide services and (2) programme partner earning incentives.

These incentives were put in place in 2012 but when the Elizabeth line December 2018 opening date was missed key performance indicators in the contract became irrelevant. The implication was that with the second component of the payment plan wiped out Transcend was left with no financial incentive to prioritise the allocation of resources to Crossrail above other projects in which its shareholders had a stake.

---

[9] Second National Audit Office Crossrail report: Completing Crossrail, 3 May 2019

The programme partner incentive arrangements were revised in February 2021 with TfL approving a supplemental agreement to the programme partner contract setting out a new mechanism for quarterly incentive arrangements. TfL did not reveal details of how this would work but it did note that "unlike the project delivery partner, the programme partner had, and continues to have, very limited scope to influence the cost and schedule outcomes of the Crossrail project". That would seem to suggest that project delivery partner Bechtel did have an impact on costs and delivery dates.

## Central figures

The National Audit Office takes great care not to apportion blame and indeed it would be unfair and overly simplistic to blame Crossrail's delay and cost rises on a few individuals. That said I believe it is important that those charged with leading and steering the development of Crossrail should be held to account. When key people are paid salaries higher than the UK Prime Minister and are in receipt of regular bonuses, drawn up in an attempt to ensure the success of the programme, surely it is reasonable to ask the question – did they perform the roles they were paid to do?

Among those in the spotlight is former Chief Executive Andrew Wolstenholme. He took the Crossrail top job in 2011 having previously spent five years in the Army before going on to manage delivery of the Heathrow Express rail link and Heathrow Terminal 5. His departure from Crossrail was announced in March 2018 at a time when the programme was said to be 'on time, on budget'.

According to Transport for London's annual report and statement of accounts for 2017-18 the project boss was paid £97,734 'compensation for loss of employment' as well as performance related pay of £160,000 (relating to 2016-17) during the year. This took his total remuneration for 2017-18, excluding pension contributions, to £736,157. At £946,396 his total remuneration for 2016-17, excluding pension, was approaching a million pounds.

I find these pay structures mesmerising. Do we pay premium rates to get people we think can deliver a project successfully? Or is it about finding someone willing to accept a 'poisoned chalice' role? Is it reasonable to pay the same rate if key goals are not met? And if not could we attract executives with a lower salary but more realistic expectations of their levers of control on a project?

In reviewing his six and a half year tenure we might pay tribute to the major engineering and tunnelling achievements realised on his watch. However, it's clear that Wolstenholme was unable to deliver Crossrail on time or on budget. After leaving Crossrail he joined BAE Systems' naval and armoured vehicles business; he was appointed to lead the maritime business with a remit that included focusing on programme schedule and cost performance. This role ended after a year. In 2021 he took up a new job at contractor Laing O'Rourke with responsibility for the delivery of technical excellence for clients, enhancing the company's digital capability and its health and safety priorities.[10]

Another key figure was Simon Wright who joined Crossrail as Programme Director in 2014 and briefly served as Chief Executive following Andrew Wolstenholme's departure. Wright, previously Project Development Director at Network Rail and prior to that Director of Infrastructure and Utilities with the Olympic Delivery Authority, does not appear to have been able to ensure Crossrail adhered to either budget or schedule.

Working alongside Wolstenholme and Wright was Crossrail Chairman Terry Morgan who moved to Crossrail in November 2009, having previously worked as Chief Executive of London Underground public private partnership contractor Tube Lines.

The PPP, under which Tube Lines was one of three consortia tasked with modernising the London Underground, is another story which

---

[10] Laing O'Rourke press release, Laing O'Rourke appoints Andrew Wolstenholme as Group Technical Director, 27 April 2021

I will not dwell on here. I think it's fair to say the model, foisted on London by central government, was neither a success in terms of what it delivered for passengers or value for money. London mayor Ken Livingstone and his successor Boris Johnson were from different ends of the political spectrum but they both wanted to be shot of the PPP arrangements as quickly as possible.

Post Tube Lines Terry Morgan was present at Crossrail and TfL panel meetings for more than a decade. Unlike other senior staff at Crossrail Ltd he attempted to stay in post once the delays and overruns emerged. However, he resigned on 5 December 2018 as both Chairman of Crossrail Ltd and HS2 Ltd following a radio interview when Morgan said he expected to lose his jobs because the Crossrail sponsors – the DfT and Transport for London – were unhappy with delays to the project.[11]

Terry Morgan placed at least some of the blame for the delays to Crossrail on TfL. During an interview with Radio 4's PM programme he said: "What most people don't realise is the rolling stock, the trains, are a Transport for London contract. I'm not responsible for rolling stock and neither are any of the executive. I personally now realise TfL were advised that this train contract was running 18 months late and have known that for at least eight months."

Morgan's comments correlate with those made by his former colleague Rob Holden, Chief Executive of Crossrail between 2009 and 2011, in a letter to *The Times* in which he described the delay to completing Crossrail as a national embarrassment which could have implications for Crossrail 2.[12] "The problems with Crossrail arise from a decision taken in 2011 to delay the procurement of new rolling stock – a decision that has affected the commissioning of those trains, the infrastructure and the all-important signalling system," wrote Holden. "The signalling was always going to

---

[11] DfT press release, Crossrail and HS2 Chairman steps down, 5 December 2018
[12] At the time TfL was seeking funding from government for Crossrail 2.

be the single biggest risk to the Crossrail project, and so it has proved."[13]

TfL subsequently rejected Holden's claim but anyway Crossrail Ltd, and by default Morgan, had been privy to the commissioning of Bombardier's Elizabeth line Class 345s for long enough. Surely then he would have had a reasonable grasp of progress at the Litchurch Lane assembly facility? And while earlier delivery of the new trains *might* have avoided delays with the stage one and two commissioning, would it really have sped up resolution of the complex signalling and software issues in the central operating section given that so much construction work was running late, impacting on the ability to run trains through unfinished stations?

After Crossrail delays and cost overruns became clear the London Assembly Transport Committee attempted to unpick what had been happening. The panel's report of 23 April 2019 recounted that in 2018, as the scale of the problems facing the programme became apparent, Crossrail's Remuneration Committee denied Chief Executive Andrew Wolstenholme, Programme Director Simon Wright and Finance Director Matthew Duncan access to their Long Term Incentive Plan payments in 2018 because of failure to meet targets.[14]

The minutes of the remuneration meeting state that Crossrail executives "reluctantly accepted" non-payment of their retention bonuses and that Wolstenholme wrote a letter to the Remuneration Committee requesting a review of its decision in March 2018 not to release his LTIP payment. According to the committee's report: "This attitude is symptomatic of a culture that, while encouraging unchecked optimism, has also encouraged a denial of responsibility."

---

[13] Libby Purves and reform of the honours systems, Letters to the Times, 4 September 2018
[14] London Assembly Transport Committee, Derailed: Getting Crossrail Back on Track, April 2019

Wolstenholme left the same month, after seven years leading the Crossrail programme, leaving Simon Wright in charge. Wright and Duncan also left the project before a new team was brought in to oversee completion.

Outside Crossrail Ltd the London Assembly Transport Committee called for the most senior employee of parent organisation Transport for London – London's transport commissioner Mike Brown – to resign following the publication of messages which appeared to show he watered down warnings to Mayor of London Sadiq Khan about the scale of problems affecting Crossrail.

The Transport Committee's report reproduces a series of email threads in which messages from TfL Head of Corporate Affairs Andy Brown, Crossrail Chief of Staff Lucy Findlay, Crossrail Chief Operating Officer Howard Smith,[15] TfL Senior Advisor to the Managing Director Sylvia Mannah and TfL Corporate Affairs Manager Stephanie Doyne noted changes to Crossrail updates prepared for Khan. In several instances Mike Brown is said to have made changes which made problems affecting the Crossrail schedule appear less serious than stated in the original text.

In one update to the Mayor, prepared in June 2018, Howard Smith was asked if he was happy with the removal of text saying there was insufficient time to complete testing required ahead of the joint trial running and trial operations from 1 October 2018. Smith responded: "Not really! Removing the statement re insufficient testing time is wrong. We see it as a critical issue that we need to note." Sylvia Mannah replied that Mike Brown was not happy including the paragraph.

Speaking in front of the London Assembly Transport Committee at City Hall on 25 April 2019 Mike Brown said: "I'm not reflecting on whether I'm fit to be in position. I believe I am, I've got full support of the Mayor, and that's the end of the issue for me." TfL announced

---

[15] Later Director of the Elizabeth Line.

in October 2019 that Mike Brown would be stepping down.[16] He was succeeded by Andy Byford in June 2020.[17]

While former senior executives of Crossrail Ltd and TfL may be a convenient lightning rod for blame, the responsibility of the Crossrail sponsors, Transport for London and the Department for Transport, in the failures should not be overlooked. Did they really have no clue how much trouble the project was in? Or was it politically convenient to eschew all knowledge until it could not be hidden any longer? It seems almost risible that Mayor of London Sadiq Khan, or Transport Secretary Chris Grayling, could claim to have been unaware of the extent of the problems facing Crossrail.

On 6 September Khan told the London Assembly that Crossrail had not informed him of the delay to opening Crossrail until 29 August. Yet TfL was told of a likely delay on 19 July. Surely the Mayor – as Chair of TfL – would have been briefed then too?

When Terry Morgan resigned as Chair of Crossrail Ltd Khan thanked him for his service but added: "For a while now I have had concerns about the effectiveness of Crossrail Ltd's governance. Not only was I angry when Crossrail Ltd informed us that the project would be delayed, but historically there has been a lack of adequate information shared by the senior Crossrail leadership with the project's joint sponsors – namely the Department for Transport and Transport for London." One wonders what, if anything, the sponsors did in response to this 'inadequate information'.

TfL subsequently noted, when Crossrail trial operations began in November 2021, that the Crossrail Board led the project with neither the DfT nor TfL being in control of the governance or construction of the line. Really? As project sponsors the Mayor, as head of TfL, or Transport Secretary, as head of the DfT, could have with minimal

---

[16] TfL press release, Mike Brown MVO set to leave TfL, 17 October 2019
[17] TfL press release, Andy Byford appointed London's new Transport Commissioner, 27 May 2020

effort acquainted themselves with the true status of the project. Whether they wanted this knowledge or not is a separate matter.

When the House of Commons Public Accounts Committee published, on 3 April 2019, *Crossrail: progress review*,[18] it provided a highly critical assessment of the sponsors' handling of the Elizabeth line and their explanations when the stage three opening was not met. According to the committee: "It is unacceptable that the Department and Crossrail Ltd are unable to identify the root causes of the programme unravelling so quickly and so disastrously." It continued: "The Department and Crossrail were unable to provide a convincing explanation of the root causes of the delays."

While taking evidence to inform their report members of the panel expressed incredulity that references in the Jacobs P-Rep report of April 2018 – before Crossrail accepted the December 2018 opening would be missed – did not attract more scrutiny. Chief Executive Mark Wild said schedules for the stations started to slip around 2016.

For all the shortcomings within the Crossrail Ltd organisation MPs were clear that the programme sponsors too had messed up. "Despite acknowledging that there were major failings in the programme, the Department and Crossrail Ltd [as part of TfL] have been unwilling to accept their responsibilities for the significant delays and cost overruns of the programme." The report went on to add: "neither are willing to identify who should be held responsible for these failures, and simply assert that there are systems failures. We are entirely unconvinced by this rationale as the Department and Crossrail Ltd were responsible for creating and managing the system that enabled these failures to occur."[19]

---

[18] House of Commons Committee of Public Accounts, Crossrail: progress review, 1 April 2019
[19] House of Commons Committee of Public Accounts, Crossrail: progress review, 1 April 2019

## Learning from mistakes

While the official announcement that the Elizabeth line would not open as planned in December 2018 was made a few months before, in August 2018, the National Audit Office found that the breakdown of contractual mechanisms – including delays and cost increases – had been clear within Crossrail Ltd for years. It said that by 2015 problems had begun to emerge and opportunities to change approach were missed.[20]

But the success of the Crossrail enterprise depended on its multiple parts, people and companies working together. One may reasonably ask how well Bechtel and Transcend lived up to their delivery partner roles. Looking at the remarkable increases in costs for some of the tier one contracts we might also question how internationally renowned construction firms missed target prices so spectacularly. Even if incentives were lacking could they not have liaised more efficiently with other contractors to minimise compensation events? In 2021 the National Audit Office found contractor performance at meeting milestones set by Crossrail Ltd continued to be low, at around 30%,[21] a woeful figure even if partly explained by schedule changes.

But it was the project delivery body, in this case Crossrail Ltd, on which so much depended. Based on the Crossrail experience it is imperative that in future rail delivery bodies make every effort to:

- Keep sponsors up to date about contract cost increases and delays
- Ensure reliable and timely access to sites for contractors
- Have a clear grasp of work that remains to be done and what this involves
- Log the status of assurance documentation for digital systems
- Wait for construction to be finished (or sufficiently advanced) before starting train testing

---

[20] Second National Audit Office Crossrail report: Completing Crossrail, 3 May 2019
[21] Third National Audit Office Crossrail report: Crossrail – a progress update, 9 July 2021

- Minimise contract interfaces when determining how contracts are packaged
- Avoid interfaces that encourage a compensation culture. Consider giving contractors responsibility for managing interfaces.
- Make delivery partners accountable for their remit.
- Retain staff until systems integration work is complete.

There is plenty here for both public and private sector organisations to take time to consider and my list of areas to focus on is in no particular order. It is clear that, despite the public presentation until 31 August 2018, Crossrail's problems were deep rooted. Many people and companies contributed to the programme's success but, unfortunately, some of those same people and companies acted in ways that saw costs rise and delays multiply. That has not only affected Crossrail but made it more difficult for other large rail programmes to proceed.

## Moving on

While examining and understanding what went wrong is important so that we can build confidence in the delivery of future rail schemes, Crossrail needed to move on. After the crisis the programme required a reboot. So who would step up to face the challenges and get the Elizabeth line open to passengers?

# 12. THE REBOOT

It was London Underground Managing Director Mark Wild, appointed as Crossrail Chief Executive in November 2018, who would steer the project to its conclusion.

The first phase of building Crossrail was about engineering, much of which was executed with aplomb. But the next phase would be about computers and code; perhaps age and experience meant the top team had not fully grasped the challenges this posed?

Wild arrived with considerable experience of railway systems. Prior to taking up the LU role in 2016 he had served as managing director of Westinghouse Signals, responsible for integration projects such as the upgrade of the Victoria line in London as well as the delivery of communications-based train control metro lines around the world including the Downtown Line in Singapore, Airport Line in Taiwan and Line 5 in Beijing. He had also served as special advisor to the Minister and Secretary of the State Government of Victoria, Australia, and been Chief Executive of Public Transport Victoria, Melbourne's integrated transport authority.

Stepping up as Crossrail chief Wild's approach was to reset rather than continue. There would be no quick fixes pledged and he refused to be pressured into making promises and committing to new opening dates. Instead, he took time to examine in detail the challenges facing the programme before setting out the way forward.

Appearing before the London Assembly Transport Committee on 9 January 2019 Wild presented a downbeat picture of Crossrail's current status. One month after the new railway had been due to open he told the committee there was still much work to do on trains, signalling, stations and tunnel fit-out. He informed the committee that all the nine new Elizabeth line stations remained unfinished.

On the overground Elizabeth line branches the picture was similarly bleak. More than seven months since TfL Rail had begun operating the stage two (of five) Crossrail service west of Paddington main works had yet to start on some of the station upgrades promised for the suburbs. The contract for work at Acton Main Line, West Ealing and Ealing Broadway stations had yet to be awarded by Network Rail.

For all that was left to do across the programme the biggest obstacle to opening the Elizabeth line remained getting the new trains to work with the multiple signalling systems fitted along the route. Mark Wild expressed confidence in the legacy Train Protection Warning System (TPWS) fitted on Network Rail infrastructure outside the Crossrail tunnels and the Trainguard communications-based train control (CBTC) system installed in the tunnels by Siemens despite the addition of what he described as "novel functions".

However, installation of signalling equipment in tunnels had taken longer than envisaged. Wild explained that the explosion of a transformer at Pudding Mill Lane in November 2017 was less significant, in terms of halting train testing, than had been presented at the time. He said signalling was not fully installed and associated software was not sufficiently mature. "If I'd known testing had started without signalling being complete we wouldn't have started," he said.

Clearly the issue of getting on-train computer systems to reliably switch between the two signalling systems was causing much head-scratching for Bombardier and Siemens; both were working together to refine the software that handled the transition to ensure no interference with other trains or tunnel systems. In a sign of how difficult this task was proving Crossrail Ltd said it hoped to be able to run two trains simultaneously within the Crossrail tunnels by mid February 2019. A milestone, yes, but still a long way from metro capability.

In May 2018 Crossrail's stage two launch had been due to see Class 345 trains running to Heathrow. But by 2019 the problem of implementing a suitably reliable Level 2 ETCS system for use in the

Heathrow tunnels had still to be resolved and there were fears it was drawing development time away from the Elizabeth line central section.

Rather than spread Bombardier resources between the two work streams a decision was taken to go all-out on commissioning the TPWS/CBTC transitions so that the Elizabeth line could open. Wild said Bombardier staff around the world were working on the issue; only when it was solved would resources be available to reconsider the Heathrow ETCS difficulty. "Heathrow at the moment is a little bit out in the long grass," he admitted. In a sign that the software team was very much a limited and finite resource he added: "We'd hoped to get it done by the end of 2019 but that entirely depends on getting Bombardier's resources out of the central section."

Despite arriving at a time of crisis for Crossrail Mark Wild's carefully chosen words in front of committee meetings consistently showed his care not to be critical of the Crossrail team before him. Still, he was clear that the strategy of letting staff go and winding down the Transport for London subsidiary ahead of the scheduled December 2018 opening date had been wrong. "In the organisation it is probably my most critical risk – I'd use the term hollowed out – the Crossrail Ltd organisation has unfortunately lost some critical mass over the last year," said Wild.

The new Crossrail boss addressed this by promoting former operations director for TfL Rail and the Elizabeth line Howard Smith to Chief Operating Officer and appointing a new Finance Director, David Hendry, along with new programme and technical directors. Crossrail's Mark Cooper was subsequently given the new role of Chief Projects Officer and Jim Crawford lured from HS2 Ltd to the newly created post of Chief Programme Officer.

By March Wild was warning that there was no chance of stage three opening during 2019, dismissing the 'autumn 2019' opening that the project sponsors had alluded to when it was announced at the end of August 2018 that the December 2018 opening could not be met.

## A new programme

With promised Elizabeth line stage three openings in 2018 and 2019 failing to materialise Mark Wild was understandably reluctant to offer new opening dates or issue promises. By the middle of 2019 though a revised schedule had been developed which envisaged the Elizabeth line opening at some point during a six month window between October 2020 and March 2021. However, Bond Street station would open later and, for now, opening dates for this as well as stage four and five services remained 'to be confirmed'.

Also still TBC, despite being a year behind schedule, was the ETCS system to allow the Class 345 train diagrams to be extended from Hayes & Harlington to Heathrow Airport. Testing had, however, by this point resumed.

A modicum of good news – the start of Elizabeth line trains from Maidenhead to Reading, previously planned as part of stage five, would now happen in December 2019 under a newly invented stage 5A. Counter-intuitively this would occur before the stage three start of services through the central tunnelled section.

To get the Elizabeth line open Crossrail Ltd now outlined four major workstreams that needed to be completed:

- Build and test the software to integrate the train operating system with three different signalling systems
- Install and test vital station systems
- Complete installation of equipment in the tunnels and test communication systems
- Once done, trial run the trains to shake out any problems and ensure the highest levels of safety and reliability once passenger services start

With work continuing at all the Elizabeth line stations, Victoria Dock and Pudding Mill Lane portals became the first Crossrail sites to be handed over to the infrastructure manager, TfL subsidiary Rail for London Infrastructure, on 28 August and 4 September, respectively. That left just 28 sites still to transfer.

In a statement to the London Stock Exchange on 8 November 2019 Crossrail Ltd's parent company Transport for London said latest projections showed a central cost forecast (including risk contingency) of approximately £15.363 billion. This was £400 million more than the funding agreed by government on 10 December 2018.

Taken together with the money provided to Network Rail to deliver the overground sections of what would become the Elizabeth line, the Stock Exchange statement indicated that the total cost of delivering Crossrail infrastructure was now expected to be between £18.2 billion and £18.45 billion. This was the third increase in the cost of Crossrail in a year.

As detailed in chapter 10 the financing package agreed by the Department for Transport, the Greater London Authority and TfL in December 2018 saw the funding envelope for the central section of Crossrail, managed by Crossrail Ltd, increase from £12.8 billion to just under £15 billion, bringing the total financing package to £17.6 billion. In July 2019 Transport Secretary Chris Grayling revealed that Network Rail costs had increased by £200 million, taking the Crossrail budget up to £17.8 billion. The Stock Exchange announcement saw the projected cost of infrastructure for the cross-London railway exceed this allocation and break the £18 billion mark.

With Mark Wild having ruled out a 2019 opening TfL's Stock Exchange statement now also scotched the possibility of a 2020 opening. Instead, it said only that the Elizabeth line would open "as soon as practically possible in 2021".

Wild said Custom House, Farringdon and Tottenham Court Road Elizabeth line stations would be complete by the end of 2019 with the fit-out of tunnels set to finish in January 2020. He added: "The two critical paths for the project remain software development for the signalling and train systems, and the complex assurance and handover process for the railway; both involve safety certification for the Elizabeth line."

Wild continued: "Crossrail Ltd will need further time to complete software development for the signalling and train systems and the safety approvals process for the railway. The trial running phase will begin at the earliest opportunity in 2020; this will be followed by testing of the operational railway to ensure it is safe and reliable."

Among the challenges of getting Crossrail finished fire systems emerged as another problem area. Following the handover of the Victoria Dock and Pudding Mill Lane portal sites to Rail for London Infrastructure, TfL noted that the Mile End shaft handover was delayed partly because of a fire alarm control panel which needed to be upgraded to meet the standards required for the railway. It went on to list 31 instances of these sub-standard fire panels installed across the rest of the Crossrail stations, shafts and portals.

Earlier in the year TfL had said that a safety critical fire system at Canary Wharf, fitted out by Canary Wharf Contractors for Canary Wharf Group and completed well ahead of the other new Elizabeth line stations, did not meet legal standards. Drawing attention to sub-standard fit-out at the Canary Wharf site, TfL then said it would need to spend more than £75 million at the station to bring safety critical systems, including those relating to fires, up to legal standards. But the revelation of fire panel problems elsewhere suggested a programme-wide issue with fire systems rather than one linked to CWG's project management.

In an effort to speed up the handover of Elizabeth line sites, and in turn to allow trial running of trains to begin, Crossrail Ltd drew up 'handover execution plans' for each station, shaft, portal and railway system. Five of the first elements to be handed over were identified as 'nursery' elements from which key lessons for handing over the later, more complicated stations, shafts and portals could be learned.

## Reading stage 5A launch

Crossrail's recently-invented stage 5A service, which would see Class 345 trains run between Paddington main line station and Reading, began on 15 December 2019.

The launch followed the announcement in March 2014 by the Crossrail sponsors that the Elizabeth line was to be extended west from Maidenhead to Reading with an intermediate stop at Twyford. At the time Rail Minister Stephen Hammond said that the Reading extension would result in cost savings from reduced infrastructure enhancements at Maidenhead and Slough with only minor works being required at Twyford and Reading to accommodate Crossrail services. The plan then was that Crossrail services would be extended to Reading in December 2019, a year after the main route through central London opened.

Instead, rather than being an extension of the full cross-London east-west/west-east service – the final stage of the Crossrail programme – for now the Reading link would be an extension of the TfL Rail branded service that MTR ran between Paddington and Hayes & Harlington.

Shiny new Class 345s at Reading at least raised awareness that, at some point, passengers would be able to board direct trains to the West End, Canary Wharf and Stratford. The new TfL service saw pay as you go ticketing extended all the way to Reading and the application of TfL fares to the route from January 2020 was expected to see the cost of adult fares from stations between Iver and Reading fall.[1]

So for now, alongside all the engineering marvels of Crossrail, running to Reading over existing lines was essentially a rebadging exercise with TfL Rail operator MTR taking over services previously run by GWR. The existing frequency of four stopping trains an hour in the peak between Paddington high level and Reading was retained – but now used Class 345s – while two trains an hour ran off-peak, as was previously the case. TfL Rail services from Reading did not call at West Ealing, Hanwell and Acton Main Line stations while

---

[1] TfL said support for Oyster smartcards would not be extended beyond West Drayton due to limitations with the system – which by then was more than 15 years old.

some fast services continued to be operated by GWR from Reading, Twyford, Maidenhead and Slough to Paddington.

## Keep your distance

By early 2020 Crossrail Ltd was committing to a summer 2021 opening for the Elizabeth line. For the first time since the delays to the programme became apparent in autumn 2018, Crossrail also shared a date for when the full, five stage Elizabeth line service would be operational. With the core route through Crossrail tunnels expected to open in summer 2021 the full joined-up network – whereupon trains to and from Shenfield, Heathrow and Reading would continue through central London tunnels – was expected to be operational a year later by 2022. This suggested that Mark Wild did not see any insurmountable issues connecting the new-build line to the national rail network once trains and signalling had been proven to work reliably in the central underground section.

No sooner had an opening date been announced than Crossrail received another curve ball – and one that it really could not have been expected to predict. The onset of the Covid-19 pandemic, and Prime Minister Boris Johnson's announcement on 23 March that people should stay at home wherever possible, would mean big problems for construction sites – many of which paused projects and furloughed staff.

On 24 March London's Transport Commissioner Mike Brown announced that TfL and Crossrail would halt all work on project sites unless there was a need to continue for operational safety reasons. "This means that work on all such projects will be temporarily suspended as soon as it is safe to do so," he said.

In a press statement Mark Wild added: "Essential and business critical work continues across the Crossrail programme but our priority is to keep our people safe and limit movement. Last week we asked everyone who could work from home to do so and put in place measures to limit movement across the programme. As a further measure we have taken the decision with our tier one contractors to temporarily stop station activity and close our sites until further notice."[2]

This predicament added another unwelcome obstacle to finishing work on site. One consolation was that with preparations for trial running so dependent on programming, much work at computers could be carried out remotely, allowing work to continue even with Covid prevention measures in place.

Having seen what had happened before when the Crossrail organisation was wound down prematurely, this time care was taken to ensure Covid did not result in the programme laying off people with the skillsets that would need to be called upon later to finish the job. Crossrail Chairman Tony Meggs said he was focused on making sure that every person working for the project, including small sub-contractors and those employed by big companies, got paid during the site shutdowns even if they were unable to come to work. It wasn't just being nice: "We want to pay them for their welfare, but also so we can restart very, very quickly. If we weren't able to keep our supply chain intact, it would take us years to build it back up again," he said.

Crossrail site activity resumed, with safety precautions in place, in May. Despite the challenges presented by Covid-19 Crossrail was able to hand over to Rail for London Infrastructure the first central section station, Custom House, along with Royal Oak Portal, North Woolwich Portal and Fisher Street shaft. (Pudding Mill Lane Portal, Victoria Dock Portal and Mile End Shaft had previously been handed over). All central section stations, except Bond Street, had now been certified and endorsed by The Railway Assurance Board (Crossrail) as ready to support trial running.

## Stage 2B takes 345s to Heathrow

On 7 May the Office of Rail and Road granted authorisation for train operator MTR to run full-length Class 345 trains in the

---

[2] Crossrail press release, Crossrail and TfL construction sites temporarily suspended, helping limit spread of coronavirus, 24 March 2020

Heathrow tunnels, to and from the airport, using the European Train Control System. This Approval to Place Into Service allowed TfL and Heathrow Airport Ltd to complete their safety validation for the first passenger services to start in June. Originally scheduled to start in May 2018, this would now be known as stage 2B of the Elizabeth line opening.

Getting 345s to Heathrow meant saying goodbye to the Siemens Class 360 Heathrow Connect units with their legacy Automatic Train Protection (ATP) system, which had previously been needed to run through the Heathrow tunnels. Operation of 345s using ETCS Level 2 on the airport branch allowed the removal of the obsolete ATP equipment from the Heathrow tunnels as soon as Heathrow Express Class 332s had been replaced by GWR Class 387 Electrostars retrofitted with ETCS.

In a further western route development, all TfL Rail services between Paddington and Reading were now being operated with full length 9-car Class 345s, replacing the 7-car 345s that were then transferred to the Shenfield line – and which in turn allowed 30-year old Class 315 units to be withdrawn.

By late summer it had become clear that Mark Wild's projected summer 2021 Elizabeth line opening would be unachievable. According to Crossrail Ltd the impact of Covid-19 had made the existing pressures on the programme more acute. "Due to a pause of physical activity on sites and significant constraints on ongoing work time has been lost, only some of which can be recovered," the company said in a statement. Explaining the latest delay it went on to add: "Construction works were halted in March as the country went into lockdown and although it has been possible to resume works at a reduced rate, in line with social distancing guidance, the virus has delayed the project's progress."

Not only had the schedule slipped again but costs were up (again) – not just on the central section but Network Rail infrastructure where a further £140 million handout saw costs approach £3 billion.

On 24 June 2020 Crossrail Ltd activated a clause in the Bond Street station main works contract with Costain/Skanska (C412) that allowed for termination of the contract following prolonged suspension of construction works – as had been the case during Covid. Crossrail brought the work in-house, appointing a new contractor, Engie, to carry out care and custody duties for the site with Crossrail Ltd overseeing completion of the remaining works required for trial running. While cost reduction was apparently not the primary purpose of changing the contractual arrangement,[3] Crossrail estimated the change would save between £20 million and £30 million. Crossrail transparency data revealed that a £19 million settlement was paid to the Costain/Skanska joint venture on 26 June.

Meanwhile, as part of a government-provided Covid-19 rescue package for Transport for London, necessitated by the collapse in passenger revenue after people were told to stay at home, agreement was reached to wind up Crossrail Ltd and integrate its functions into parent organisation TfL. The move was said to include the governance, oversight and "actions necessary to provide greater confidence in the timely and value for money" delivery of the project.

On 1 October came the announcement that Crossrail Ltd board meetings were no more (it met for the last time on 17 September) with governance of the project transferred to TfL. According to Crossrail this move aimed to simplify responsibilities with a single TfL Elizabeth Line Delivery Group composed of senior members of TfL, London Underground and Crossrail Ltd, under the chairmanship of the capital's Transport Commissioner Andy Byford. However the executive team, led by Mark Wild (who now reported directly to Byford), remained in place and several of the board's non-executive directors continued to provide TfL with advice in relation to Crossrail.

---

[3] Third National Audit Office Crossrail report: Crossrail – a progress update, 9 July 2021

High-level oversight of the final stages of Crossrail was to be provided by a special purpose committee of the TfL Board – the Elizabeth Line Committee. This would meet in public every eight weeks and include members of the TfL Board; it would also be attended by a special representative from the DfT as joint sponsor of the project. This continued the more hands-on government approach set out in the May TfL bailout conditions and formalised in the Crossrail Transition Action Plan which was submitted to the DfT as part of this Extraordinary Funding and Financing Agreement.

In return for six months of Covid-related funding the government told the Mayor and TfL which transport projects could and could not proceed. Regarding Crossrail 2 a DfT settlement letter included an instruction to bring "an orderly end to consultancy work as soon as possible". However, the letter also said the "DfT will support such safeguarding activity for this project as required".

By the end of 2020 Farringdon became the first major station (setting aside the much more straightforward Custom House) to achieve substantial completion. Achieving this T-12 (12 weeks from handover) milestone allowed main contractor BAM/Ferrovial/Kier to start demobilising and enabled Crossrail to begin the process of handing the station over to infrastructure manager Rail for London. As 2021 began all the shafts and portals had been transferred to the infrastructure manager but Paddington, Tottenham Court Road, Woolwich, Abbey Wood, Liverpool Street, Canary Wharf, Whitechapel and Bond Street had yet to be signed off. Difficulties recruiting specialist staff to complete the mass of certification and assurance documentation required for the new railway did not help progress. One consequence of the delay to opening the Elizabeth line was that Crossrail was now facing competition from HS2 Ltd to attract those with the technical skills required.

Tottenham Court Road and Paddington reached T-12 status on 12 and 22 February respectively. Welcome signs of progress could also be seen with overground Crossrail schemes – Network Rail opened a new steel and glass ticket hall at Acton Main Line (18 March) with a new ticket hall at Ealing Broadway opening on 27 May. Under what Crossrail Ltd described as 'first phase works' Broadway gained a new glass frontage, a larger ticket hall under a timber-ceilinged canopy, and a longer gateline. Meanwhile, Grade II listed Hanwell station was removed from the Heritage at Risk Register by Historic England following extensive renovation work by Transport for London.

In reality the handover of Paddington took longer than the nominal 12 weeks so by the summer Woolwich had succeeded Custom House, Farringdon and Tottenham Court Road in being passed over to the infrastructure manager.

By the autumn these had been followed by Liverpool Street, Paddington and Whitechapel. Canary Wharf would be next: Crossrail Project Representative Jacobs reported that the sign-off of Canary Wharf had been hampered by "continued discoveries of [redacted] and non-compliant material/installation". It said issues had been found with fire systems and concern remained that further works requiring long-lead items might be necessary, adding to the schedule pressure.[4] Network Rail opened new glass and steel ticket halls at Southall (26 August) and Hayes & Harlington (14 September).

## Running the trains

By now it will be clear how much work remained to do, across the programme, well after the supposed stage three December 2018 start date had been and gone. But while it remained important to address the incomplete work at stations and other sites, getting trains running through the newly built tunnels was a priority.

The start point was to embark on a period of 'dynamic testing' that would identify and fix any software bugs in the train control system and make sure everything worked as planned. Test results would be recorded, analysed and checked to produce evidence the system was safe and met an acceptable level of performance.

---

[4] Crossrail P-Rep Project Status Report 153 – Period 4, 20 August 2021

Much of this work took place away from the new London railway. The Crossrail Integration Facility in Chippenham, Wiltshire, provided an off-site environment in which to test the critical train control systems. Chippenham acted as a giant laboratory; engineers were able to request specific equipment they needed from Siemens' different divisions around the world and then bring it together to simulate the systems that would be used to control Elizabeth line trains.

The tests enabled signalling software and data from Bombardier and Siemens to be combined and any anomalies discovered before being tested on the railway. Each new version of the train and signalling software increased functionality and the range of dynamic testing that could be undertaken. The outcome of testing conducted at the Crossrail Integration Facility provided confidence in the functionality and behaviour of the software. Once this had been proven off site a software update would be installed on the central railway section to be put through its paces during dynamic testing. The cycle would then repeat; as issues were discovered programmers updated the software and prepared the next software release.

As a result of this process Crossrail was ready to start testing in the Elizabeth line tunnels and main dynamic testing began in January 2019, initially with a single train. A second train was then introduced, allowing tests to take place in each of the twin Crossrail running tunnels. The next step was to move to close-headway, multi-train testing in 'integrated' mode; this involved testing the full signalling system in the central section with trains running at line speed.

At this point it was becoming evident just how important software would be for the Elizabeth line. Crossrail recognised that it was going to need additional time to complete programming for the signalling and train systems and the safety approvals process for the railway. Not only was much of the important activity going on at the Crossrail Integration Facility in Chippenham but Siemens' programming team in Germany was also involved in testing the software. The power to complete the new railway had shifted from the civil engineers to the software engineers.

## Target trial running

The goal for Wild and the Crossrail team was to begin 'trial running'. This was the process under which trains would be tested through the new Crossrail tunnels and the signalling and automatic train control system could be put through its paces. Through trial running engineers would build reliability and flush out any issues with systems and signalling software.

In June 2019 Mark Wild and the Crossrail top team had thought trial running would start by March 2020 with trial running estimated to take between nine and 12 months. Unfortunately, tests by the Siemens team in Germany identified missing functionality and defects in the central section software which had been slated to become the first release capable of supporting trial running. Another version of the software would be required although, once again, only after full testing had taken place would it be clear if the problems had been resolved. The commitment to a summer 2021 opening revealed that trial running would not begin before the autumn.

Again, it is hard to overstate the scale of this debugging process; this was not half a dozen programmers in an office. Mark Wild explained: "In the world of railway signalling, there are not many people in the world who are not here doing this job. The software is being built in 20 facilities by about 1,500 people all over the world at this very minute. It is probably the biggest software push in railways at this moment in the world."

Although by this stage the signalling transitions between CBTC tunnels and Network Rail interfaces were largely proven, according to Wild, it was not until December 2020 the precursor stage to trial running was achieved when Crossrail announced the start of 'system integration dynamic testing' (SIDT) which it described as the enabling phase for trial running. The start of SIDT would see the number of test trains in the new tunnels increase from four to eight.

System integration dynamic testing involved a forensic examination of complex systems to trial scenarios as close to operational conditions as possible. Signalling and control of the railway was transferred to Crossrail's new control centre at Romford, bringing those based at the facility into the test programme for the first time. Drivers operated a maximum of eight trains, simulating a timetabled service across the Elizabeth line central section.

By spring 2021 Mark Wild was able to tell the London Assembly "the positive news is that we have not found any issues that we do not already have solutions for in future software releases". But it would take several months more for the Crossrail team to be ready to start the proper trial running phase on 10 May. This involved integrated trials of the railway to demonstrate that the Elizabeth line was safe and reliable and that it met the capacity and performance requirements needed to move to the final stage before opening the railway, known as 'trial operations'.

Shortly before trial running began, the pre-existing Great Western Main Line and Great Eastern Main Line systems were fully linked with the new Elizabeth line central section to form an operational railway ready for trains to run across the entire route. Not only did this streamline the logistics of completing the railway – making it easier to move full length Class 345s across the capital from Old Oak Common depot in the west for introduction into service between Liverpool Street high level and Shenfield – it meant stage four and stage five through-running services were now possible.

In another sign of pieces slotting into place Transport for London's Rail for London Infrastructure arm was formally designated infrastructure manager for the central section under ROGS – the Railways and Other Guided Transport Systems (Safety) Regulations 2006. This gave TfL the ability to allow the infrastructure to be used for operating trains.

During trial running Crossrail steadily ramped up the number of trains running in the 42 kilometres of tunnels that had been built below London to allow the railway and supporting systems to be operated as close as possible to the future timetable. This was initially with a four train per hour service in each direction – some way off the 24 that would be needed to operate the full Elizabeth line service.

Indeed, once trial running started it stopped soon after. The programme was paused for 18 days for a planned engineering blockade, from 21 June to 12 July, that would address a long list of snags in tunnels and stations which could not be rectified while trains were operational. Trial running frequencies increased to 8 trains per hour on 7 June and, following the blockade, were upped to 12. This was significant as it would be the initial frequency of Elizabeth line trains when they eventually began passenger service.

Once again the software engineers were at the heart of this process. Systems were fitted and ready to go but reliability was an issue. Large scale software updates were scheduled periodically, including during a Christmas 2021-New Year shutdown to address the seemingly ever-present glitches that continued to be picked up during testing.

## Time running out

By mid December 2021 it had been three years since Elizabeth line stage three Crossrail services had been scheduled to start. Chief Executive Mark Wild had taken time to understand and unravel the problems that had delayed the programme. He had been cautious about committing to new dates and making promises which, like those of his predecessors, might end up broken.

And yet he had now had three years leading the project and counting. A first half of 2022 opening commitment had been given, after missing the 2021 start. Wild's reputation – and that of his boss Andy Byford – were on the line. After more delays could they finally get the Elizabeth line open?

When I wrote my December 2021 Crossrail column for *Modern Railways* I was conscious there were barely six months left in which to achieve that goal. Mark Wild's latest update to the London

Assembly revealed plenty left to do: Canary Wharf and Bond Street stations had yet to be handed over to infrastructure manager Rail for London. Another train control software update would be required. While this was being commissioned improvements would need to be made to the tunnel ventilation system. And, to be ready to carry passengers, the trial operations phase needed to get underway as a matter of urgency.

Distinctive yellow escalator surrounds at Canary Wharf. Despite being the first station built it was among the last to be ready for handover to the infrastructure manager.

# 13. THE OPENING

For anyone planning to deliver a new cross-city railway, hooking up existing lines either side with new tunnelled infrastructure, the benefits of a phased opening are obvious: break it down into key components and test each of these work before bringing them together. Far more likely you'll get the infrastructure working this way than trying to switch everything on all at once.

And so, when the go-ahead for the project was given in 2010, Crossrail – the Elizabeth line naming decision was announced in 2016 – was to open in five stages. Stage one would see the service operator take over existing commuter services between Liverpool Street and Shenfield which would eventually be integrated into the cross-London route. Similarly, stage two would transfer the Paddington to Heathrow service, previously operating as Heathrow Connect, to the Crossrail operator.

Stage three would mark the big step forward – the opening of the Crossrail tunnels with trains running between the new Paddington Crossrail station and Abbey Wood. But the western and eastern overground branches would remain separate until stage four, which would join up the Shenfield route with the central operating section. Finally, stage five would add on the routes west of Paddington to complete the full Crossrail service.

## Commissioning plan at start of Crossrail programme

| Stage one | May 2017 | Liverpool Street (high level) to Shenfield |
|---|---|---|
| Stage two | May 2018 | Paddington (high level) to Heathrow Airport |
| Stage three | December 2018 | Paddington (Crossrail) to Abbey Wood<br>(Crossrail service through central tunnels launches) |
| Stage four | May 2019 | Paddington (Crossrail) to Abbey Wood and Shenfield |
| Stage five | December 2019 | Heathrow and Maidenhead to Abbey Wood and Shenfield<br>(Full Crossrail service operating) |

We have already looked at some parts of this phased opening but, what follows, is the full Crossrail opening sequence. It turned out to be somewhat different to the original plan.

## Stage one – 22 June 2017

As we saw in chapter 9, on Thursday 22 June 2017 members of the public were given their first opportunity to ride on Elizabeth line trains. The first Class 345 had been due to enter service with the bi-annual national rail timetable change on 21 May but driver testing took a little longer than planned.

Initially the Aventra-based sets operated on the Liverpool Street high level to Shenfield route, branded as TfL Rail. During the remainder of 2017 the majority of services on the TfL Rail Liverpool Street to Shenfield route switched to being operated by new trains – although some of the older trains, by now over 30 years old, remained in service until stage three.

Crossrail Operations Director Howard Smith told the Transport for London board[1] that the launch was delayed from May because further releases of software for the Train Control and Management System were required to address issues such as passenger and driver displays. While these did not prevent training and testing they were deemed unacceptable for passenger service. This need to update the software held up final safety and regulatory sign-offs.

Smith also expressed concerns about the stage two launch, due in less than a year, which would require the Class 345s to operate with the European Train Control System on the Heathrow spur. Testing between the ETCS on-train kit and the track-side systems was planned but meeting the next deadline was expected to prove tough. Smith told the board that "experience elsewhere in Europe demonstrates that detailed incompatibility issues often arise".

---

[1] Transport for London Board, 19 July 2017, Elizabeth Line Operations and Transition Update

Laboratory testing involving both Alstom[2] and Bombardier systems and physical tests of the trains with the track equipment were scheduled for the second half of 2017.

As preparations continued for the stage two launch Heathrow Airport announced a small but significant change to the Elizabeth line route.

Crossrail plans had long envisaged trains serving what is now the Heathrow Central stop and terminating at or starting from the Terminal 4 station. However, the start of trains through the central operating section would now see six Elizabeth line trains an hour serving Heathrow in each direction with four starting/terminating at Terminal 4 and two starting/terminating at Heathrow Terminal 5.

The agreement by Heathrow, Transport for London and the Department for Transport meant that, together with Piccadilly line services and four Heathrow Express trains an hour to Heathrow, services to the airport were predicted to increase from 18 per hour to at least 22 with the stage three opening. A commitment was also given to undertake a joint feasibility study looking at how to deliver a further two Elizabeth line trains per hour to Terminal 5.

## Stage 2A – 20 May 2018

As we heard earlier, it was apparent well before the May 2018 launch date that despite the Class 345 trains being fitted with ETCS, and the Heathrow tunnels being wired up with ETCS, that compatibility was going to be a problem. Expectations were doused and, when the timetable change arrived, Crossrail was able to introduce a limited number of Class 345 diagrams from Paddington high level, but initially venturing no further west than Hayes & Harlington. The older, but Heathrow tunnel compatible, Heathrow Connect Class 360 sets would be retained by the new Crossrail

---

[2] In this instance Alstom was the supplier of track mounted equipment used with ETCS

operator and remain in service a little longer.[3] Technically Crossrail could claim it had met its stage two target with the transfer of services to the new operator. However, rather than being a big switch on moment with all-new train sets it proved to be a phased introduction with 360s remaining in use with TfL Rail until 2020.[4] Subsequently the labels 2A and 2B were attached to refer to the initial and then the full stage two Crossrail service.

## Stage 5A – 15 December 2019

The biannual national rail timetable change on 15 December 2019 saw Crossrail operator MTR take over from GWR the running of most stopping services between Paddington high level and Reading under the interim TfL Rail brand.

As described in chapter 12 stage 5A represented an enhancement not envisaged when construction of Crossrail began; extending from Maidenhead to Reading had been a safeguarded aspiration until it was added to the programme in 2014 by the Conservative-LibDem coalition government.

With Crossrail struggling to meet opening schedules extending MTR's services beyond Maidenhead to Reading may have looked like an attempt to claw back credibility. However, the 5A launch was primarily a hard-nosed financial decision – Transport for London expected to collect an extra £190 million in ticket revenue over a three-year period as a result of the 5A service.

## Stage 2B – 7 May 2020

As set out in chapter 12, approval from the Office of Rail and Road allowed train operator MTR to run full-length Class 345 trains in the tunnels to Heathrow Airport using the European Train Control System Level 2 signalling system. This followed

---

[3] Class 360s used the now obsolete ATP system which was fitted on the Heathrow route
[4] Modern Railways, October 2020, page 85

extensive software development to resolve compatibility issues between trackside equipment and kit fitted on board the Aventra trains.

## Commissioning strategy revamped – July 2021

Chapter 9 recounts in detail how Crossrail went on to miss the stage three opening date of December 2018 as the delivery schedule slipped and costs multiplied. Official acknowledgement in August 2018 that the new tunnels and stations were not ready marked the beginning of what turned out to be a protracted period of refocusing the programme.

Initially, despite the delays and difficulties, the assumption was that the remaining delivery dates for stages three, four and five would be moved forward accordingly. In early 2020 Crossrail Ltd was talking about a summer 2021 stage three opening date and projected the full opening of the cross-London route to take place a year later. There was little, if any, mention of stage four and with hindsight it appears that Mark Wild's team were by this point considering if there might be a better commissioning strategy than the one they had inherited.

To recap, with MTR having already taken over the running of suburban routes from Paddington and Liverpool Street, stage three would see the Elizabeth line open and then stages four and five would bolt on the existing MTR operations, to Shenfield and Reading/Heathrow respectively, to provide the full cross-London rail experience. In considering alternative commissioning strategies, two key factors were at play. Firstly, might there be an approach that would better mitigate risk? After all, getting systems working on time and integrating technology had already proved problematic. Secondly, with Transport for London finances facing a triple squeeze by government, the impact of Covid on ridership and the hole in the balance sheet caused by the delay to the start of Elizabeth line revenue-earning services, could a different phasing get more paying customers using the line than would be the case with the original steps from stage three to four and five?

Papers for an Elizabeth line committee meeting in July 2021 revealed that Mark Wild's team thought they could do better.[5] Stage three would go ahead as planned with a service – introducing the Elizabeth line branding – operating through the new tunnels between Paddington and Abbey Wood. No trains would run through the Royal Oak or Pudding Mill Lane portals with MTR continuing to run standalone services to Reading and Shenfield from high level Paddington and Liverpool Street platforms.

This stage three would be followed by a new stage 5B which would add the proposed stage four through service from Paddington to Shenfield but would simultaneously divert services from Reading and Heathrow through the new Elizabeth line tunnels to Abbey Wood. Crucially, while these two routes would overlap, they could be operated entirely separately, bringing in most of the benefits of the final Crossrail configuration, now referred to as stage 5C, but without all of the technical complexities.

To better understand this consider the following: stage four would have required a large number of trains to reverse at Paddington – not ideal given that commissioning the automatic reversing functionality[6] was some way off. Under the 5B plan this requirement would be halved by running two overlapping railways, significantly increasing resilience. Existing Reading and Shenfield timetables would work – services would simply be extended into the tunnels – so no national rail timetable changeover date complications would arise. Driver training schedules would be under less pressure because drivers need only be trained on one set of national rail route procedures (beyond Royal Oak or Pudding Mill Lane), not both.

From a financial point of view this intermediate stage 5B would allow passengers from the Shenfield route to travel direct to the West End and, similarly, those from Heathrow and Reading line stations would be connected directly to central London and Canary

---

[5] Elizabeth line committee meeting, 15 July 2021
[6] See chapter 8

Wharf. Also, no need to maintain a TfL Rail brand for the suburban routes – instead Elizabeth line branding could be applied across the railway, enhancing the status of the entire route and providing a public relations boost as well as possibly justifying more lucrative advertising rates.

That left 5C, formerly stage five, as the final Crossrail service pattern. The two overlapping rail routes would become one, allowing Shenfield trains to continue to western destinations and vice versa. True cross-London travel would be possible without a change of trains. But after the big changes of stages 3 and 5B this final phase might not even be noticed by passengers.

## Crossrail commissioning as it happened

| | Details | Start of Class 345 services |
|---|---|---|
| Stage one | Liverpool Street high level to Shenfield | June 2017 |
| Stage 2A | Paddington high level to Hayes & Harlington | May 2018 |
| Stage 5A | Paddington high level to Reading | December 2019 |
| Stage 2B | Paddington high level to Heathrow | May 2020 |
| Stage three | Paddington low level to Abbey Wood (introduction of Elizabeth line brand – without Bond Street) | May 2022 |
| Stage three | Bond Street station open | October 2022 |
| Stage 5B minus | Shenfield to Paddington low level and Abbey Wood to Heathrow/Reading (two overlapping but separate routes). No auto-reverse at Paddington. | November 2022 |
| Stage 5C | Full, integrated, cross-London service. Auto-reverse available at Paddington. | May 2023 |

## Ramping up to stage three

On Saturday 20 November 2021 Crossrail began trial operations, the final step before passenger services could begin. The shift from trial running to trial operations marked a move from a closed system overseen by the technical teams to a railway where members of the public would be invited in to be involved in final testing.

Trial operations included exercises to ensure the safety and reliability of the railway for public use and to test the robustness of timetables. More than 150 scenarios were acted out to ensure the railway was ready for passenger service. Systems and processes needed to work effectively and train and station staff be able to respond confidently to any incidents, including customers being unwell or signal failures that could leave a train stranded underground. To test cross-party working London Underground, MTR, Network Rail and the emergency services were all involved during the running of the different test situations.

To meet its self-imposed trial operations deadline TfL split the process into two phases – the first phase was characterised mainly by an increase in the frequency of test trains. The start of phase two a few weeks later saw activity intensify with invited guests and volunteers taking part in station evacuations to street level as well as being escorted through Crossrail tunnels and along track to the nearest station after simulated train breakdowns.

Mass testing events involving members of the public travelling on the Elizabeth line were completed successfully over the weekend of 11-12 March 2022 and confidence was now high that the railway was nearly ready.

The next step was for the Elizabeth line to undergo a short period of shadow running to ensure that the initial planned timetables worked as expected and operating procedures were sufficiently robust. Crossrail would operate as it would run on opening but without fare-paying passengers. Having successfully completed this phase Crossrail applied to the Office of Rail and Road for the final permissions required to open the railway. Testing by Transport for London staff continued until the necessary documentation was finalised. With everything in place all that was needed now was a green light from the regulator.

This was confirmed on 13 May with the ORR declaring that the routeway and all new stations, with the exception of Bond Street, met the requirements for passenger use. Under UK law no new or upgraded infrastructure or rolling stock can be put into use on or as part of Britain's rail system unless the ORR has provided an 'interoperability authorisation for the placing in service' to ensure it meets appropriate requirements.

The ORR approved the Elizabeth line Class 345 fleet in 2020 and subsequently authorised the Global System for Mobile Communications-Railway (GSM-R) for use on the new infrastructure. This delivers digital, secure and dependable communications between drivers and signallers.

The regulator also approved the track access contract between Rail for London (Infrastructure) Ltd, the infrastructure manager for the Crossrail Central Operating Section, and train operator MTR Corporation (Crossrail) Ltd. If, in the future, another train operator wants to use the Elizabeth line it will require a track access contract that has been reviewed and approved by the ORR.

## Stage three – 24 May 2022

The much anticipated opening of the Elizabeth line, London's new east-west underground railway, took place on Tuesday 24 May 2022.

Station gates on the central section opened to the public early on Tuesday 24 May when Elizabeth line staff welcomed the first members of the public, some of whom had queued through the night in order to catch the first westbound service from Abbey Wood or eastbound train from Paddington. There was plenty of enthusiasm from those keen to sample the new east-west railway with many dressed or decorated in purple. Across the capital London landmarks – parks, bridges and buildings – were illuminated in purple to draw attention to the fact that the Elizabeth line was now open.

Upon opening trains ran every five minutes from 0630 to 2300, Monday to Saturday, between the new Paddington Crossrail station and Abbey Wood, providing passengers with 12 trains an hour on the core Elizabeth line. With the exception of Queen Elizabeth's platinum jubilee weekend in early June TfL was able to use Sundays to organise extended engineering possessions to update train and signalling software.

Despite the opening of the new Elizabeth line tunnels the full Crossrail service had yet to arrive. Any passengers wishing to travel end to end along the Elizabeth line, from Reading to Shenfield, would have to change twice (at Paddington and Liverpool Street). For the time being the Elizabeth line operated as three separate routes.

With the start of stage three services passengers gained access to eight vast new stations at Paddington, Tottenham Court Road, Farringdon, Whitechapel, Liverpool Street, Canary Wharf, Woolwich and Custom House. Elizabeth line stations offered passengers new connections with direct access from platforms to Barbican, Moorgate and Soho plus the ability to reach Gatwick, Stansted and Southend airports with one change of train.

From opening all Elizabeth line stations were staffed from first to last train. Step-free access from street to train was available at all the new stations from Paddington to Woolwich although TfL said that at the rebuilt Abbey Wood station some passengers might want to use a manual boarding ramp. At Custom House wheelchair users were advised to board the fifth carriage of Elizabeth line trains to ensure level access.

In 2016 Boris Johnson, then Mayor of London, announced that Crossrail would be known as the Elizabeth line in honour of the long-serving monarch. Despite opening at least three and a half years late the Elizabeth line opened in time for the extended platinum jubilee weekend marking 70 years on the throne for HRH Queen Elizabeth II. The Queen never rode on an Elizabeth line train but on 17 May, as one of her last public engagements before her death in September 2022, she visited the new Paddington station, experiencing the completed new railway which bears her name and which will help ensure she is remembered for many years to come. The Queen previously visited Bond Street station during construction in 2016.

After the struggles to pay for the escalating cost of Crossrail the opening of the Elizabeth line – with subsequent ticket sales – switched on a much-needed new revenue stream for Transport for London as owner and infrastructure operator of the new tunnels.

TfL was happy to report that in the first week since opening more than one million journeys were made along the central section of the Elizabeth line between Paddington and Abbey Wood. Add in the surface routes – operating as TfL Rail prior to being rebranded when the central section opened on 24 May – and more than two million journeys were made along the entire route from Reading and Heathrow in the west to Shenfield and Abbey Wood in the east.

Indeed, the first days of passenger service went rather smoothly. On day one the tripping of an over-eager fire sensor at Paddington prompted a train evacuation at Tottenham Court Road but the matter was quickly dealt with and services continued uninterrupted. Day two saw the central section achieve an impressive 100% reliability. In the build-up to the Elizabeth line opening reliability had been a key concern but the ELR200 on-train signalling software, commissioned at Easter, generally performed well. Further software updates would later be rolled out to enable a full through-London service to operate at peak frequencies.

Elizabeth line operating hours between Paddington and Abbey Wood were extended from 5 September with Monday to Saturday services beginning an hour earlier – at 0530 rather than 0630.

## Bond Street – 24 October 2022

On Monday 24 October 2022, following ORR approval, Bond Street became the final station to open as part of the Crossrail programme. New ticket halls (at either end of the two platforms) now provide direct access to London's Oxford Street and Mayfair. This station alone, described in detail in chapter 5, represents an infrastructure investment of well over half a billion pounds.

## Stage 5B minus – 6 November 2022

Stage 5B of the Elizabeth line opened on 6 November 2022 when, for the first time, national rail services starting outside the capital ran direct to London's central shopping and Docklands business districts.

The change replaced the previous three separate Elizabeth line services (west of Paddington, central tunnels, east of Liverpool Street) with two overlapping services: Reading/Heathrow to Abbey Wood and Shenfield to Paddington.

As a result many users of the Elizabeth line central section, which opened six months previously, no longer needed to change train to reach their destination. TfL said business travellers would, for the first time, be able to use a single train to travel between Heathrow Airport and Canary Wharf with journeys taking as little as 45 minutes. By making a Thameslink connection at Farringdon travel between Heathrow and Gatwick or Luton airports now required only two trains. Stage 5B created many other direct travel opportunities – passengers from Reading, Slough, Ilford and Romford, for example, could now reach the City and West End without changing at a main line terminus.

Aside from the new through journey opportunities the introduction of overlapping train diagrams provided a major uplift in train frequencies for existing Elizabeth line users. Services in the central section between Paddington and Whitechapel increased from 12 trains per hour to up to 22 trains per hour at peak times and 16 trains per hour off-peak. This brought frequencies close to the

24 trains per hour peak promised for the final Elizabeth line service configuration.

In addition to the stage 5B launch (technically '5B minus' given the yet-to-be-completed auto-reverse system at Westbourne Park sidings) another development brought the full Crossrail vision closer to realisation. Regular Sunday services began on 6 November, turning the Elizabeth line into a full seven day a week operation, although planned weekend closures on overground sections of the network, as well as Christmas and New Year engineering possessions, meant some interruption to this.

While direct journey opportunities to and from Heathrow were created by the stage 5B launch, discount fares to the airport ended following a condition imposed by government as part of its latest funding agreement to prop up Transport for London after the Covid-19 pandemic.

As of Sunday 4 September, all Elizabeth line and London Underground fares for journeys that went through Zone 1 and started or ended at Heathrow Airport were charged at peak rate, adding £2 to the cost of an adult pay-as-you-go fare outside peak hours. Journeys to the airport avoiding Zone 1 continued to have an off-peak fare, while journeys ending at stations before Heathrow, such as Hayes & Harlington, were not affected by the change.

TfL also revealed that to regulate the frequency of services coming into Paddington Elizabeth line station some trains would be held outside the Royal Oak Portal in west London for a few minutes before being routed into the Crossrail eastbound tunnel. This delay was said to be factored into timetables and was announced to passengers on board affected trains.

Ahead of the stage 5B opening TfL announced that step-free access was now available at all Elizabeth line platforms following the completion of Network Rail's main works programme. Ilford and Romford stations on the Elizabeth line's Shenfield branch became the latest upgraded stations to be commissioned. Work at Ilford had been delayed following the discovery of flaws in the concrete slab which supported the station; Network Rail brought in contractor Murphy to undertake repairs.

Following the launch of stage 5B December 2022 saw two milestones relating to rolling stock. For years operator MTR had been running a mix of 7 and 9-car Class 345 Aventras along Elizabeth line surface routes with the fleet composition dictated by short platforms at Paddington and Liverpool Street high level stations. With platform lengthening work now complete at Liverpool Street,[7] and with trains no longer having to terminate at Paddington main line station, MTR was able to convert the three remaining 7-car trains to 9-car full-length units. This meant, five and a half years after the introduction of the first Class 345 into passenger service, the Elizabeth line could operate a homogeneous fleet of 70 9-car trains, each 205 metres long and able to carry up to 1,500 passengers per train set.

With the Crossrail rolling stock deployment complete, December also saw the retirement of the final Class 315, previously used by MTR on the Liverpool Street to Shenfield route. Built in York for British Rail, retirement of these trains meant the end for the fleet after 42 years passenger service across London and East Anglia.

## Final push

2023 began with a focus on resolving minor issues affecting the recently launched stage 5B service and ensuring infrastructure and systems were resilient enough to allow 5C Elizabeth line trains to run through central London every 150 seconds.

Despite this activity the trajectory for completing Crossrail was now clear and structures put in place for delivery could increasingly be stood down with confidence that they were no longer required. TfL described the Crossrail project moving to become a 'close out organisation' from 16 January.[8] Chief Programme Officer Jim

---

[7] See chapter 5
[8] Transport for London Board, 7 December 2022, Elizabeth Line Operations and Further Opening Stages

Crawford stepped down with responsibility transferring to Crossrail Close Out Director Kim Kapur, reporting to TfL Elizabeth line Director Howard Smith.

Issues being addressed at this stage included platform screen door seals becoming detached when caught by large luggage (particularly at Liverpool Street and Paddington stations where many passengers change for the high level stations), and overhead electric line failures between Paddington and Reading – Network Rail devised a programme of interventions designed to increase resilience. A glitch that prevented trains automatically switching between signalling systems at Stratford was temporarily overcome by reducing the speed of trains over a short section of track while Siemens developed a software fix.

Optimising software was another key focus area. Updated SCADA (Supervisory Control and Data Acquisition) communications software was commissioned at the end of October 2022 and the ELR300 signalling software upgrade, which would allow the operation of 24 trains per hour at peak times along the central section of railway, was deployed at Christmas. A further signalling upgrade, known as ELR400, was due to be commissioned during Easter 2023 and coincided with the completion of all station contracts.

The Elizabeth line central operating section between Paddington and Abbey Wood was closed from Friday 7 April to Monday 10 April with the blockade used to commission upgrades to both the ELR400 train control software and the line's communications system over the Easter weekend. Elizabeth line Director Howard Smith said the upgrade would remove 21 operational restrictions, close 194 minor software issues and deliver 13 further functional improvements. This included the long anticipated auto-reverse facility at Westbourne Park, west of Paddington, which went live after work had been split into two sub-projects (tunnels and Westbourne Park sidings).

## Stage 5C – 21 May 2023

Stage 5C launched on 21 May 2023, replacing the previous two overlapping service diagrams with end-to-end, east-to-west,

west-to-east trains across London including off-peak Shenfield-Heathrow journeys. Crossrail – the ability to travel from one side of the capital to the other without changing trains – had arrived.

The final Elizabeth line increment saw the railway move to a peak service of 24 trains per hour in the central section, an increase from 22. Initially 16 services per hour were due to run in each direction off peak. Trains to and from Heathrow became more frequent and, for the first time, direct Shenfield-Heathrow services were introduced. Reading saw a small increase in peak trains with the Elizabeth line taking on a handful of diagrams that were previously operated by GWR.

Perhaps the biggest improvement for those passengers already using the Elizabeth line to travel into Paddington was the removal of any significant pause for trains between Paddington and Acton Main Line stations. These had been in place since November 2022 to regulate train arrivals at the Royal Oak tunnel portal but improvements to train control systems, together with the new timetable, meant these planned stops were no longer needed. That in turn meant faster runs between Reading and Paddington and reduced journey times for Elizabeth line users travelling from the west into central London and Docklands.

## Opening sequence complete

The opening of stage three in May 2022 was a purple moment; at last the public and the media could see the new railway they had been promised: new trains, new tunnels, new stations. As transit systems go there was plenty of wow factor.

In many ways stage 5C, one year later, was more significant. It marked the end of the vast Crossrail undertaking (the end of the beginning, anyway). Goals had been achieved, the train service was fully operational, people and resources that had for so long been occupied by many interlinked Crossrail projects had finally been released.

Yet, travelling along the Elizabeth line on 21 May 2023, I was struck by the absence of fanfare. Unless you happened to be heading for

Heathrow from Shenfield the change to the final service increment and the new direct journey opportunities were hardly obvious. In a similar vein the end of pauses outside portals and the uptick in train frequencies were subtle changes and probably not enough to alert most people to the fact that today was the day the Crossrail programme had completed.

The truth was, a year after stage three launched, the Elizabeth line had already established itself as an integral part of London's transport system. Already the novelty of the new railway was wearing off and instead it was increasingly forming part of the daily routines of thousands of people.

Crossrail's success seems assured. One year after the Elizabeth line opened Transport for London reported that more than 150 million customer journeys had been made on the line during the year with around 600,000 journeys being made each weekday.[9] Tottenham Court Road saw over 100,000 additional journeys passing through the station each year, more than doubling usage since May 2022. A YouGov survey showed that nearly half of Londoners had used the new railway since it opened.

Initial figures from TfL suggested that across the entire Elizabeth line 19% of demand represented journeys that would previously have been taken on London Underground lines and 4% on the Docklands Light Railway. While it was inevitable that the new railway would to some extent cannibalise other routes and modes, the figures appeared to suggest many new journeys were being generated and that most people choosing to use the Elizabeth line would not have considered travelling by train before.

After the many technical challenges of commissioning Crossrail the Elizabeth line is regularly proving to be the most reliable railway in Britain on the basis of the moving annual average performance

---

[9] Transport for London press release, 22 May 2023, Full peak Elizabeth line timetable introduced as railway celebrates remarkable success in its first year

measure. One example: in Period 12 (5 February to 4 March 2023) the only train operators recording greater reliability than the Elizabeth line during this period were Merseyrail, London Overground and Greater Anglia.

One would expect Crossrail to transform travel in the capital but its impact shows up well beyond the M25. According to Office of Rail and Road figures quoted by TfL covering the period between October and December 2022, one in six of all rail journeys in Britain was made on the Elizabeth line. Out of all the train operators monitored by the ORR the Elizabeth line was ranked second place for number of journeys (62.2 million), just behind Govia Thameslink Railway's 68.8 million (which covers a sprawling network including Southern and Gatwick Express services running in and out of London Victoria). Within a year of operation, ahead of the figures coming in for the full 5C increment, the Elizabeth line was already one of the most used railways in the United Kingdom.

Farringdon's integrated ticket hall on Cowcross Street with access to Elizabeth line platforms to the left of the departure screens.

# 14. THE OPERATOR

Trains, tracks and tunnels do not, by themselves, make a fully functioning railway. The equipment and the technology is only as good as the force that knits it together – in this case the Elizabeth line operator.

Running trains – reliably and on time – is the operator's primary remit. But it also plays a wider role in the passenger experience of the railway: the operator manages most of the stations along the route, it runs ticket offices, dispatches trains and ensures trains are cleaned. In short, whether a passenger has a pleasant journey on the Elizabeth line largely comes down to the system operator.

First impressions count – clean trains, helpful staff – but to ensure the Elizabeth line functions seamlessly requires the operator to pay close attention to its interfaces with other organisations. At Old Oak Common depot the operator's fleet team works closely with train supplier and maintainer Alstom. Delivering a reliable service is only possible when the infrastructure works and so that requires close liaison with Network Rail and TfL's Rail for London arm. The operator also needs efficient back office logistics – its responsibilities have included the recruitment and training of staff, the attainment of safety certification and the provision of resources to support the introduction of the new rolling stock, depot and station facilities as well as associated testing and commissioning activities.

## Making concessions

The operator's remit – what it is expected to do and how – is set out by Transport for London. TfL is in charge of the Elizabeth line service but does not run it; the day-to-day workings are outsourced to the operator via a concession agreement. TfL retains the revenue risk for services (as it also does for the London Overground and Docklands Light Railway). The thinking is that MTR will be able to focus on those areas it can control and the economic uncertainties that upset many national rail franchise projections pre-pandemic and distracted management attention can be avoided.

271

Freed from worries about how many fares are sold the operator is remunerated through a fixed concession payment that is adjusted according to performance against a suite of metrics, many of them linked to train punctuality. The concessionaire is responsible for the operation of ticket offices and ticket vending machines and for undertaking revenue protection activities.

TfL has plenty of experience using this model given its time running the Docklands Light Railway and the London Overground system; the Elizabeth line operating concession is more sophisticated but is essentially an Overground mark two.

## Operational expert

Throughout the genesis of Crossrail, when it comes to Elizabeth line operational matters, one name keeps cropping up. Howard Smith, TfL's Director of the Elizabeth line at the time of writing, was appointed Operations Director for Crossrail Ltd in 2013.

The appointment was with a view to getting Elizabeth line services up and running; moving beyond infrastructure construction and tooling up TfL to run a new train service. Smith knows more about the programme than most – his involvement with Crossrail dates back to 2004 – prior to joining the project team he was on the TfL/Department for Transport joint sponsor board.

Howard Smith survived the turmoil of the later stages of the Crossrail programme because his operational knowledge of metro railways is unrivalled. Prior to taking up his Crossrail role in 2013 he spent nine years as Chief Operating Officer of TfL Rail, responsible for the transformation of London Overground, and he spent a further six years before that with TfL as Director of the Docklands Light Railway.

## Contract award

On 30 July 2014 TfL awarded a £1.4 billion contract to Hong Kong Metro operator MTR to run Crossrail trains from 2015 until at least 2023. The deal included an option for a two year extension.

MTR, which at the time was running London Overground services in partnership with Deutsche Bahn-owned Arriva, saw off competition from Arriva, National Express and Go-Ahead/Keolis to land the contract.

The Crossrail concession began on 31 May 2015 with MTR initially operating trains between Liverpool Street high level and Shenfield, taking over the stopping services previously run by Abellio Greater Anglia. Prior to the arrival of new Elizabeth line rolling stock it leased 44 of the Class 315 EMUs used on the route from Eversholt Rail with these maintained by the Greater Anglia franchisee at Ilford. These 4-car units were refurbished and progressively replaced with the new Class 345 trains ordered from Bombardier.

Train operators can inevitably expect some changes to services to occur during a contract term but the scale of upheaval planned during the MTR Crossrail contract, later MTR Elizabeth line (MTREL), was arguably unprecedented. In July 2014 TfL set out the following schedule for the new concession holder:

## Crossrail train operator schedule

| Stage 0 | May 2015 | Crossrail train operating company takes on Liverpool Street (high level) to Shenfield services operated by the existing Class 315 fleet |
|---------|----------|-------------------------------------------------------------------------------------------------------------------------------------|
| Stage 1 | May 2017 | Introduction of the first Class 345 unit into passenger service |
| Stage 2 | May 2018 | Start of four trains per hour between Paddington (high level) and Heathrow Terminal 4, replacing Heathrow Connect and part of Great Western inner suburban service |
| Stage 3 | December 2018 | Formal introduction of the first Crossrail services through the central operating section between Paddington (Crossrail) and Abbey Wood |
| Stage 4 | May 2019 | Introduction of trains from Shenfield to Paddington (Crossrail) through central operating section |
| Stage 5 | December 2019 | Completion of the Crossrail project. Introduction of all Crossrail services including through London to/from Reading and Heathrow Terminal 4 |

Later on in the Crossrail programme there was a reluctance to accept that any rephasing might be necessary so it is interesting to note that, as part of the concession bidding process, TfL asked prospective operators to price in a series of options which would allow some of these dates to change. Bidders were requested to calculate the cost of delivering stage five at the same time as stage four, an option which if exercised would see all Crossrail services operating by May 2019, about six months ahead of the plan. Shortlisted firms were also asked to price in a one year delay to stages three, four and five, an option which could mean a full Crossrail service was not in place until the end of 2020.

Shortly before contract signature TfL said it would include these prices in the concession agreement, providing price certainty should it need to invoke these options following contract award. Other options bidders were asked to price included extending train services to Reading – an option subsequently taken-up, the cost of reducing or increasing Crossrail train kilometres, and a price for retaining a proportion of the Class 315 fleet to provide residual peak services to and from Liverpool Street.

TfL said the £1.4 billion contract figure was the nominal value of concession payments payable to the operator over the eight year concession term before performance adjustments including bonuses and penalties. As part of the concession agreement MTREL was required to secure its obligations with an on demand bank bond of £15 million, increasing to £25 million at the beginning of stage three, and a parent company guarantee of £80 million.

Although the construction of Crossrail is a remarkable endeavour, it reached an end point with completion of the new railway. In contrast the system operator's role is relentless – every day it is contracted to provide approximately 700 timetabled services a day, a peak timetable frequency of 24 trains per hour and an estimated 11.4 million train service kilometres per year. MTREL was expected to employ around 1,100 staff on the Elizabeth line – most of these new roles – including nearly 400 train drivers.

## More about MTR

MTR's bid to run Crossrail services drew on extensive experience operating rail services around the world. At contract award the Hong Kong-listed company ran the metro, light rail and Airport Express services in Hong Kong and was the developer and manager of significant rail-related property interests. MTR was the lead shareholder in the operation of the Melbourne train system in Australia, ran the Stockholm Metro concession in Sweden and had an international consulting business. In China it operated Beijing Metro Lines 4 and 14 and the Daxing Line extension, Shenzhen Metro Longhua Line and Hangzhou Metro Line 1.

In addition to experience working with Transport for London on the Overground network, MTR had an understanding of TfL operating practice from Jay Walder, TfL Managing Director, Finance and Planning until 2007. He became Chief Executive Officer of MTR Corporation in 2012 before leaving in August 2014.[1] Walder was photographed at a signing ceremony for the Crossrail train operating concession in one of his final public roles for the company.[2]

Although MTR's experience and credentials for running a metro system were extensive there have been questions asked about the company following the Chinese government's intervention in the Hong Kong Special Administrative Region and crackdown on protests in 2019 and 2020. Following the partial privatisation and public listing of MTR on the Hong Kong Stock Exchange in October 2000, the Hong Kong SAR Government owns about 75% of the company.[3] Although there remain many differences between Hong Kong and mainland China the SAR is, ultimately, answerable to Beijing. Given tensions between the UK and China the existence of a

---

[1] First Report by the Independent Board Committee regarding the Hong Kong Section of the Guangzhou-Shenzhen-Hong Kong Express Rail Link, MTR website, 16 July 2014
[2] MTR Signs Concession Agreement to Operate Crossrail Train Service in London, Railway News, 30 July 2014
[3] MTR website, Investor's Information

MTREL has entered into station access agreements with train
operating companies for surface stations including Shenfield.

Chinese state-backed organisation running London's flagship
transport service may to some appear less innocuous than it did
when the contract was awarded.

MTR Corporation says that, despite its majority ownership by the
Hong Kong SAR Government, the company is independently
managed on commercial principles, financially independent and
does not rely on any government subsidy.[4]

## The birth of a brand

The commissioning of Crossrail, and the phased introduction of
train services, is covered in chapter 13. When MTR took over the
running of the existing western and eastern overground services
(which would eventually be connected) in May 2015 they were
operated under the holding brand 'TfL Rail'. Understandably, the
Crossrail sponsors did not want to introduce Crossrail-specific

---

[4] MTR website, Investor's Information

branding on fleets of aged train sets and a service which did not run across the capital.

Unfortunately this holding brand, with its dark blue roundel, would be a fixture of London's transport network for longer than anticipated. According to the opening schedule it would be needed for three and a half years when through-London services were scheduled to begin. But delays to the opening date, when they became apparent, significantly extended the lifespan of the TfL Rail brand.

Crossrail, as we know now, would never become an operating brand, an official line name. For years there had been rumours that, when the new railway opened, it would not be called Crossrail. In 2013 Crossrail boss Andrew Wolstenholme told me: "It's not my job to change the name but I do know the Mayor is rumoured to quite like the name the Elizabeth line."[5] In February 2016, during a visit by the Queen to the under-construction station at Bond Street, Mayor of London Boris Johnson confirmed the change: Crossrail would from this point forward be known as the Elizabeth line in honour of Her Royal Highness.[6]

The decision was said to be partly linked to the Queen's long association with UK transport; she was the first reigning monarch to travel on the London Underground in 1969; in 1977 she opened Heathrow Central station (Terminals 1 2 3) on the Piccadilly Underground line; and, following the 7 July 2005 terrorist attack on the London Underground network, the Queen unveiled a plaque at Aldgate station in 2010 remembering the lives of the 52 people who had died.

## Station management

The Elizabeth line operator manages most overground stations under lease with Network Rail. Where new stations on the central operating section interface directly with London Underground

[5] Modern Railways Crossrail Supplement 2013
[6] TfL press release, Crossrail to become the Elizabeth line in honour of Her Majesty the Queen, 23 February 2016

stations LU is responsible for managing the Elizabeth line facilities; that includes Bond Street, Tottenham Court Road, Farringdon, Liverpool Street and Whitechapel (the Crossrail operator is required to provide staff for Elizabeth line platforms). TfL's Rail for London subsidiary manages the new Paddington (low-level), Canary Wharf, Custom House and Woolwich Elizabeth line stations. The concession holder has entered into station access agreements with the franchised train operating companies which manage Shenfield, Slough and Maidenhead stations as well as Heathrow Airport Ltd, which owns and manages the three Heathrow stations used by Crossrail.

## Romford control role

Elizabeth line train care is based at the Old Oak Common depot and infrastructure maintenance facility at Plumstead. But control of the new railway is co-ordinated from Romford.

Network Rail's Romford Rail Operating Centre and Liverpool Street Integrated Electronic Control Centre control train movements into London Liverpool Street with the second floor at Romford fitted out to manage the communications-based train control (CBTC) system used to co-ordinate the passage of Elizabeth line trains from Westbourne Park to Abbey Wood and Pudding Mill Lane. Crossrail trains further west fall under the control of Network Rail's Thames Valley Signalling Centre ROC at Didcot while, at launch, the Pudding Mill Lane to Shenfield section of the Elizabeth line was the responsibility of the Liverpool Street IECC.[7] Although control of the eastern section of line is expected to transfer to Romford the initial arrangement meant the central section control (at Romford) was based at a centre east of the eastern control centre (Liverpool Street).

Romford's role extends beyond train and signalling control. Staff from TfL, Network Rail, the Elizabeth line operator and Alstom work side-by-side to deal with the day-to-day challenges of running a railway, responding to any incidents or other disruption along the route as well as handling routine but vital tasks such as co-ordinating

---

[7] Signalling Challenge for Crossrail, Rail UK, July 2020

passenger information, managing ventilation and dealing with enquiries from passengers.

## Track access

The organisation of Britain's railways is poised for a shake-up but at the time of writing most train operators pay track access fees for their use of rail lines with the charging regime overseen by the Office of Rail and Road. Within this framework the Elizabeth line operator is required to pay fees to Network Rail but also Heathrow Airport Ltd, which owns the rail infrastructure from Airport/Stockley Junction where the Heathrow branch diverges from the Great Western main line. As well as being used by Elizabeth line services this branch, opened in 1998, is also used by the airport's Heathrow Express.[8]

However, although the Elizabeth line operator agrees and pays track access fees the no-risk concession structure means TfL stands behind these agreements to reimburse track access fees.

Ahead of the Elizabeth line opening MTR was required to negotiate track access fees with the infrastructure owners. During this process Heathrow Airport Ltd argued that Crossrail users should make a contribution towards the cost of building the Heathrow spur. While MTREL would recoup the cost of any increased track access fees from TfL, fare-payers would ultimately end up paying the bill.

In May 2016 the Office of Rail and Road ruled that Heathrow Airport Ltd could charge MTR Crossrail for access to its infrastructure but should not be allowed to recoup historical costs of building the spur in the 1990s.[9] The airport subsequently launched a judicial review of the decision, a move which threatened to delay track access agreements and the introduction of Elizabeth line stage two services to the airport, scheduled to start in May 2018.

---

[8] The Heathrow Express contract is due to end in 2028
[9] Office of Rail and Road, Charging Framework for the Heathrow Spur, May 2016

Following a three day High Court hearing in February 2017, Mr Justice Ouseley dismissed Heathrow Airport Ltd's application and upheld the 2016 decision of the ORR.

The ruling was announced in May and, in a lengthy discussion of legal issues, the judge noted the many commitments given by Heathrow to a range of parties over many years to support construction of Crossrail. He said: "There is a continuing dispute over terms of [track] access, before the ORR. Were the ORR not to be arbiter, if the Regulations did not apply, HAL's commercial power would be untrammelled, putting years of work at risk."[10]

## MTR extension approved

When the eight year Crossrail operating contract was awarded, going live in May 2015, that must have seemed ample time to cover the multiple stage commissioning and bedding down of new Elizabeth line services. The reality, as we have seen, was very different to TfL's 2014 schedule and it proved a struggle to finish the new railway before MTR's operating contract expired.

In October 2022 Transport for London revealed that it would activate the two year extension provided for in the original contract, maintaining MTR operation until May 2025. A decision to take up the two year option was taken shortly before 28 August, the last date by which TfL was entitled to exercise the option to extend.[11]

## 2032 and beyond

On 19 November 2024 TfL announced that it would award a new concession, let on similar terms to the previous one, to GTS Rail Operations, a joint venture between Go Ahead Group, Tokyo Metro and Japan's Sumitomo Corporation.[12] The new contract will run

---

[10] Royal Courts of Justice, case number CO/4518/2016, 25 May 2017
[11] TfL Finance Committee meeting, 6 October 2022
[12] See also Class 345 sale and leaseback in chapter 7

from 25 May 2025 for seven years with an option to extend for up to two additional years.

TfL previously shortlisted four bidders[13] for the second generation Elizabeth line ops contract. As well as GTS Rail Operations bids were submitted by Arriva UK, a joint venture of First Group and Keolis, and incumbent operator MTR Corporation which, by the time its contract ends, will have run Elizabeth line trains services, and their precursors, for a decade. Staff employed by MTREL were set to transfer to the new operator under TUPE[14] regulations.

Announcing the concession award TfL noted that MTR had 'overseen numerous milestones on the railway including the introduction of new rolling stock, the opening of the Central Operating Section, and consistently high customer satisfaction scores'. Under the new contract TfL will continue to set fares. It said revenue generated by ticket sales would be reinvested in improving the network and there would be no immediate changes to service levels or train times.

## Operational experience

Running a train service sounds so simple and yet, even with TfL harbouring the revenue risk for ticket sales, wide-ranging demands have been placed on the Elizabeth line operator, which has far-reaching responsibilities. The concession holder has had to balance a vast mobilisation effort, as the different stages of Crossrail have come on-stream, with the day-to-day running of a metro service; initially inheriting sub-optimal trains and stations but now charged with delivering a service which lives up to the expectations of the billions of pounds invested in the route. Like the best hotels perhaps the operator's success can be measured by how unaware most Elizabeth line passengers will be of the efforts that have gone

---

[13] TfL press release, Bidders shortlisted in the process to find the next operator of the Elizabeth line, 16 February 2024

[14] Transfer of Undertakings (Protection of Employment) Regulations 2006

into making sure they experience a smooth and unremarkable journey.

Behind the Elizabeth line operator stands TfL. While the operator translates TfL policy into action it is TfL that must formulate, refine and articulate that policy. According to Howard Smith the ambition of the TfL operations team is to create a railway that sets the benchmark in European metro operations and aspires to be at least as good as the best anywhere on the continent.[15] Long after construction of the Crossrail tunnels and stations is complete it is this operational model which will underpin the experience of travelling on the Elizabeth line.

[15] Modern Railways, Crossrail operation feature, March 2016

# 15. THE FUTURE

This book tells the story of how Crossrail came into existence, culminating in the full Elizabeth line stage 5C service, introduced in May 2023, at which point Crossrail, the project, was completed.

Yet this completion marked the start of a new journey – one which will see the Elizabeth line evolve and adapt in response to changes in London, the rail system and the way we live and travel. Post-opening the railway has already changed – USB sockets have been fitted on board the Class 345 trains and the roll-out of mobile coverage in Elizabeth line underground stations (4G) and tunnels (5G) was completed in 2024. What else could the future have in store?

## Rethinking routes

Go back far enough and you will find proposals for Crossrail routes extending into suburbs all around London. Part of the genius of the Elizabeth line is that it links together existing lines to produce a public transport offering that is more than the sum of its parts.

With the troublesome technicalities of joining up new and existing railways now figured out Crossrail is just that – connected. Physically there is relatively little to prevent Elizabeth line routes outside the central section changing or being added to. Many observers have noted how Crossrail is unevenly balanced; two full routes to the east and only one to the west – a limited number of trains use the Heathrow spur – with many trains currently turned round at Paddington. Might it be possible to extend these to new destinations to extract greater value from the investment in Crossrail?

In 2004 proposals were unveiled for an alternative Crossrail scheme, dubbed Superlink. Like the Elizabeth line, Superlink would have seen new rail lines built between Heathrow Airport and Canary Wharf, but with Piccadilly and Victoria Tube line interchanges and no station at Whitechapel. However it would also connect with existing suburban lines to allow trains to run from Guildford, Basingstoke, Reading, Northampton, Cambridge, Stansted Airport, Ipswich, Southend and Tilbury through the capital.

Behind the plans were some rail industry heavyweights including Chris Stokes, involved in the development of Crossrail plans,[1] Terence Jenner and John Prideaux, credited with establishing the route for the Channel Tunnel Rail Link between Stratford and St Pancras.

Superlink's promoters said that while their scheme would cost over £2 billion more than Crossrail, it would carry four times as many passengers. By building Superlink in phases the idea was that the funding gap for the scheme would be lower than that facing Crossrail and two-thirds of the total project cost could be recouped through fares.

The response to the proposals from the authorities, including Mayor of London Ken Livingstone, was less than enthusiastic. This was not necessarily because Superlink did not have merit but, having worked hard to secure government backing for Crossrail there was a fear that re-examining options for a cross-London railway might cast doubt on the existing scheme, potentially delaying construction and providing Treasury sceptics with an opportunity to cancel the project.

---

[1] Chris Stokes, London will soon be unimaginable without the Elizabeth Line, https://cogitamus.co.uk/chris-stokes/london-will-soon-be-unimaginable-without-the-elizabeth-line

## Superlink's five point plan

| 1 | Crossrail was planned to run only to Terminal 4 at Heathrow. Superlink suggested that it be extended through Terminal 5 and integrated with the Airtrack project.[2] A short tunnel at Staines would allow Crossrail to serve Woking, Basingstoke and Guildford, while relieving congestion on trains into Waterloo. |
|---|---|
| 2 | Trains from Northampton, Milton Keynes and Watford, which currently run into Euston, should be diverted into the Crossrail tunnels west of Paddington. Besides making journeys easier and increasing capacity, this would relieve congestion at Euston on the Victoria and Northern Lines. Crossrail planned to turn 14 trains from east London around empty at Paddington which Superlink said would waste half the capacity of the new railway. |
| 3 | Crossrail plans originally envisaged running trains only as far west as Maidenhead. Superlink called for an extension to Reading. |
| 4 | Superlink proposed a new branch from Canary Wharf to Stansted, with through trains to Cambridge. This would relieve pressure on the M11 motorway and support development of Stansted as an alternative airport to Heathrow. |
| 5 | The organisation proposed a second branch in the east to Shenfield to increase capacity from the growth hotspots of south Essex. It said a connection at Barking on to the Tilbury branch would support development in the Thames Gateway. |

## West Coast connection

Although Crossrail saw off the challenger, Superlink's proposals informed development of the railway that was built. Today's Elizabeth line trains *do* serve Heathrow Terminal 5 and Reading even if some of the proposed destinations further afield remain beyond their reach.

But one part of the Superlink route that did gain traction was the idea of extending Crossrail services to run on the West Coast Main

---

[2] Heathrow Airport's owner at the time, BAA, dropped the scheme, which had faced opposition from Transport Secretary Philip Hammond, in April 2011

Line. Linking the Elizabeth line with the route for commuter trains into London Euston would require construction of a relatively short connecting stretch of track, or chord, at Old Oak Common and would potentially give Hertfordshire towns including Watford, Hemel Hempstead, Berkhamsted and Tring direct trains to London's West End, City and beyond, avoiding the need to change at Euston for connecting Underground services and typically lopping 15 minutes off journey times.

In May 2014 a meeting of Transport for London's Finance & Policy Committee heard that the Crossrail-West Coast Main Line link was one of four key issues pertaining to the capital that had been raised by David Higgins' HS2 Plus report.[3] This stated: "The view expressed by the DfT, HS2 Ltd and Network Rail is that the emerging option to deliver a better, more integrated Euston station also depends on the construction of the Crossrail West Coast Main Line link by 2024 so as to enable some of the existing train services to be removed from Euston station during construction of HS2."

By September Transport Secretary Patrick McLoughlin had announced a feasibility study to examine how Crossrail could be linked to the West Coast Main Line.[4] High Speed 2 was the driving factor here; at the time the ambition was that Euston station would have been rebuilt and expanded ready for the launch of High Speed 2 phase one services in 2026. Set against the disruption this would entail the option to reroute commuter services away from Euston, into Crossrail tunnels, looked attractive. Studies produced for the Mayor of London and Network Rail had previously calculated that a connection between Crossrail and the West Coast Main Line at

---

[3] HS2 Plus: a report by David Higgins, 17 March 2014
[4] UK government press release, government launches study into potential Crossrail extension, 7 August 2014. Patrick McLoughlin said: "I have asked HS2 Ltd to work closely with the Crossrail sponsors to look at extending Crossrail services to key destinations in Hertfordshire. Not only would this be a huge boost to passengers and the local economy, it would also provide flexibility when building HS2 into Euston, making sure we create a lasting legacy for the station."

Old Common would cost between £436 million and £489 million. Looking back at my notes I'm struck by how likely it seemed at the time that the green light would be given to construct the chord.

A decade on and these proposals seem to have been forgotten although, at the time of writing, plans to redevelop Euston remain in flux. When the Crossrail-West Coast link cropped up in a conversation I had with Howard Smith in late 2023 the Elizabeth line director was dubious about the prospects of this materialising.

## HS2: a new role for Crossrail

What is certain is that High Speed 2 will increase the importance of the Elizabeth line. When High Speed 2 opens it will initially – and probably for several years at least – start in the south at Old Oak Common, in west London. The first HS2 services are expected to run between Birmingham Curzon Street and Old Oak Common in London at some point from 2029 to 2033.[5] The Elizabeth line will gain a new station at Old Oak Common between Acton Main Line and Paddington. HS2 Ltd has agreed to build platforms and turnbacks to support this and changes to signalling will be required between Old Oak Common and Westbourne Park to enable the extended service.[6]

When the High Speed 2 plans were drawn up the Crossrail station at Old Oak Common would have acted as an interchange between HS2 and Great Western rail services. But the uncertainty concerning HS2 plans for Euston significantly changes the role of this new Crossrail station: the Elizabeth line station will now form the gateway for transporting passengers arriving on High Speed 2 onwards into central London.

Old Oak Common looks set to be a remarkable station. With six platforms for high speed services and eight for conventional trains, plus new adjacent stations on the London Overground network

---

[5] HS2 Ltd website
[6] Howard Smith, Modern Railways feature, March 2016

proposed nearby, it will be the best-connected and largest new railway station ever built in the UK. But 'connected' is right – for all the regeneration plans for the local area Old Oak will not be a destination for most of the initial HS2 passengers, at least not in the short term, so the Elizabeth line's role in transporting them onwards to where they want to go is key.

Has Crossrail been designed to do this? In a word, no. For all the discussions and debates during the development of the programme, the painstakingly refined legislation that became the Crossrail Act, providing an HS2 shuttle service to and from central London, was never part of the plan. After all, HS2 passengers for London would travel to Euston; Transport for London had spent more time considering the importance of Crossrail 2 at Euston to accommodate this influx of people than looking at what would happen if train-loads of long distance passengers disembarked at Old Oak and promptly picked up an Elizabeth line train.

Clearly there is going to be an impact on the Crossrail route. In a letter to Transport Secretary Mark Harper Mayor of London Sadiq Khan warned that by 2030 the eastbound sections of the route from Reading and Heathrow to Paddington would be "nearing capacity" and HS2 passengers would be boarding "already busy trains".[7] Strange to think that within 10 years of the full Crossrail service being delivered parts of the route could be operating close to capacity. Khan warned that crowd control measures could be required at Old Oak Common to avoid congestion on stairs, escalators and platforms.

Confirmation that the government is delaying delivery of Euston High Speed 2 station – in addition to the delays that were already known to have affected the HS2 schedule – has forced TfL to draw up new plans for the Elizabeth line to mitigate the demand now

---

[7] Khan letter to Harper, quoted in Evening Standard Online, 10 March 2023

expected when HS2 opens. Based on the anticipated High Speed 2 launch service of three HS2 trains an hour arriving at Old Oak Common, TfL expects 52,858 Elizabeth line passengers between Old Oak and Paddington going east in the morning peak (0700 to 0959) and 48,985 going west between Old Oak and Paddington in the evening (1600 to1859).[8]

What this means is that where before TfL had expected 12 trains per hour to call at Old Oak Common, effectively adding a stop to existing services to and from the west, that is now expected to increase to 20 or even 24 trains per hour,[9] probably by extending the services which currently terminate and auto-reverse at Paddington to Old Oak Common.

But, remember, that's not the Crossrail service that was planned so long ago and a significant uplift in services exceeds the capability of the existing fleet. TfL says more trains are needed: "Without them, there is likely to be insufficient capacity on the Elizabeth line for those looking to travel on High Speed 2, as they will need to use the Elizabeth line to travel to and from central London until HS2 is extended to Euston station in the 2040s."[10]

In June 2024 Alstom was awarded a £370 million contract to supply 10 new 9-car Aventra trains for the Elizabeth line along with associated maintenance until 2046.[11] The new trains, which will take the total fleet size to 80 sets and account for £220.5 million of the order value, are funded by the Department for Transport with Transport for London committing to a long-term Class 345 fleet maintenance contract. As well as preparing for enhanced Elizabeth

---

[8] Freedom of Information request, TfL Planning for HS2, published 16 May 2023
[9] Commissioner's report, TfL Board, 7 June 2023
[10] TfL press release, Annual budget for 2023/24 shows TfL set to deliver operating surplus, 22 March 2023
[11] Alstom press release, Alstom signs a €430 million contract for 10 Aventra trains with associated maintenance for the Elizabeth line in London, 14 June 2024

line service levels once Old Oak Common station opens the follow-on order also provides much needed work for Alstom's Derby Litchurch Lane site.

## The first Elizabeth line upgrade

Even before the arrival of HS2 Elizabeth line passenger numbers were increasing. As 2023 advanced the records for the number of daily and weekly passenger journeys were being broken every two to three weeks. By the end of the year TfL was reporting that 4.7 million journeys had been made on the Elizabeth line during one week with a Thursday just before Christmas recording 777,000 journeys. After having recorded more than 150 million passenger journeys in 2022/23, the first full year of operation of the Elizabeth line (following the stage three commissioning), 210 million passenger journeys were made on the Elizabeth line in 2023/24 (following completion of the Crossrail programme and the stage five commissioning).[12]

With usage of the Elizabeth line expected to continue growing *and* additional substantial demand likely to materialise when HS2 opens, the passenger carrying capacity of the line will be increased. One option is to lengthen trains; the 9-car Class 345s are capable of extension to 11 carriages given the 240 metre length of the new central station platforms. This would, however, require additional platform edge doors; when the Elizabeth line opened these were installed along 200 metres of the platform lengths so the remaining 40 metres of each platform would need to be kitted out. But a more significant hurdle would be the requirement to lengthen platforms at the Heathrow stations. Unlike the stations purpose built for Crossrail these do not have passive provision for 11-car trains and significant engineering work would be required.

The upshot is that any lengthening of trains is a medium term capacity project, unlikely before the 2030s. The first major Elizabeth line upgrade will therefore be an increase in train frequencies west of

---

[12] TfL press release, The Elizabeth line continues to transform travel in London on its two-year anniversary, 24 May 2024

Paddington, extending services that currently stop/start at Paddington/ Westbourne Park to serve Heathrow and Old Oak Common. The Crossrail signalling system has been designed to allow an increase in peak train frequency from 24 to 30 services an hour, potentially allowing a 25% increase in passenger capacity.

## New stations

Old Oak Common Elizabeth line station is happening but plans for other new Crossrail stations have been drawn up with others expected to emerge.

One proposal often talked about is to add a station for City Airport in London's Docklands. The Elizabeth line passes 300 metres to the south of the airport and Crossrail planning appears to have failed to anticipate the expansion of the airport; if plans for the line were drawn up today would trains pass by an airport serving more than three million passengers a year[13] as well as ignoring the opportunity to provide a direct public transport connection to and from Heathrow? While discussions have taken place about a Crossrail station at the airport,[14] and senior staff at the airport have indicated the airport would contribute to the cost, no firm plans have been brought forward.

On the other side of London there is plenty of evidence of planning for a new Kensal Portabello station. For more than ten years the Royal Borough of Kensington and Chelsea has lobbied for an additional Crossrail station to be built at the former Kensal Gasworks site, located between Paddington and Old Oak Common. Making the case for Kensal Portabello station the council has stated that North Kensington is poorly connected to central London by public transport and it contains wards that are among the most deprived in the UK.

During construction of Crossrail this proposal was regularly discussed, a series of studies carried out and former Mayor of London Boris Johnson was sympathetic to the plans. Ultimately the

---

[13] Civil Aviation Authority figures for 2022
[14] London City Airport in Talks with TfL about Crossrail station, New Civil Engineer, 21 May 2019

inconvenience of longer journey times for most Elizabeth line users was viewed as outweighing the benefits to local people of the additional station.

The plans drawn up by the council would have seen services currently terminating at Paddington/Westbourne Park extended to a new turnback at Kensal Portabello station. However, as HS2 proposals were developed it emerged that nearly all trains would need to run as far as Old Oak Common. The London borough has responded by drawing up plans for a new passing loop that would see a Kensal Portabello island station positioned at the centre of an enhanced track layout, allowing the station to be included or omitted from train calling patterns as timetables require. Kensington and Chelsea council has said it would be prepared to contribute towards the cost of the additional track through planning contributions. However, the cost of this looks likely to be substantial.

At the time of writing the council appears to have successfully lobbied HS2 Ltd to ensure passive provision for a Kensal Portabello station is retained.[15]

More recently plans have been drawn up for a new Elizabeth line station at Twyford Gardens to support future housing development.

Demand for the station, which would be located at the western end of the Crossrail route between Reading and Twyford, may not immediately appear obvious. However, the project was backed by property developer Berkeley, which is working with Wokingham Borough Council to deliver around 2,500 new homes and Berkeley, as we have seen, was integral to funding and delivering the Elizabeth line station at Woolwich.

Berkeley said Twyford Gardens station would relieve pressure on the existing Twyford station, which would remain open, and undertook feasibility work with rail consultants SLC Rail and Systra to show trains to the station could be timetabled without adversely affecting

---

[15] Royal Borough of Kensington and Chelsea, https://www.rbkc.gov.uk/media/document/high-speed-2---full-council-decision-on-petitioning

existing Elizabeth line services. It said it would be possible, at least in engineering terms, to locate a new four platform station near to a widened Waltham Road bridge; Ruscombe junction, which is a key piece of nearby rail infrastructure, would be unaffected.

Since setting out and consulting on these plans Berkeley has stepped back from the new station proposal to concentrate instead on enhancements to the current Twyford station. It says a new station remains an option but making improvements to the existing station, in particular new car parking and improved access, are likely to be more achievable.[16]

## After dark

Night Tube services on the London Underground began in August 2016 with a Night Overground route (using the former East London line) added in 2017. While Covid-19 saw services suspended, by August 2023 the Night Tube operated on Friday and Saturday nights on the Central, Jubilee, Northern, Piccadilly and Victoria lines alongside the Night Overground.

Could the Elizabeth line gain an all-night service? Currently the railway broadly operates the same hours of service provided across the Underground. However, to make through-the-night trains a reality would require significant changes to maintenance on the surface sections of the route where Network Rail practice is to undertake major work overnight at weekends.

Passenger numbers, and the post-Covid recovery in travel patterns, will dictate whether Elizabeth line trains through the night are viable. Following the death of the Queen TfL ran a partial Elizabeth line service through the night on the 16, 17 and 18 September 2022, allowing mourners to travel into London to attend the Queen lying in state. A limited all-night service (no trains to and from Heathrow) operated on New Year's Eve at the end of 2023.

---

[16] Berkeley, Castle End Gardens, https://castleendgardens.co.uk/improving-local-infrastructure/

## Beyond Abbey Wood

In November 2004 it was decided that Crossrail's south eastern branch would terminate at Abbey Wood. The key factor in the decision not to extend Crossrail to Ebbsfleet was that operating the existing North Kent rail services, and a new interlinking Crossrail service, could not be guaranteed without four-tracking the line between Slade Green and Dartford.[17]

The case for extending to Ebbsfleet might not seem so obvious as extending Crossrail from Maidenhead to Reading but the route would connect the Elizabeth line to key towns in north Kent. These include Dartford and Gravesend as well as the High Speed 1 infrastructure at Ebbsfleet with its fast 'Javelin' services to Kent and potentially international trains.

Almost immediately after the Crossrail Bill was submitted to parliament in 2005 safeguarding directions were issued to protect the Ebbsfleet route (which extends east of Ebbsfleet to Hoo Junction sidings) and to resolve the operational difficulties that led to this section of railway being dropped from the Bill. However, these directions were deemed insufficient to support four-tracking, and the work required at Hoo Junction, leading to further consultation work being undertaken.

Revised safeguarding directions were announced in October 2009 following a consultation on draft directions which closed on 19 December 2008. Making the announcement transport minister Sadiq Khan said the measure would future-proof Crossrail.

Although not figuring in the Crossrail construction programme local authorities have continued to push for the Ebbsfleet extension to be built. The Crossrail to Ebbsfleet Partnership (C2E) was set up in 2016 to lobby for the project to get the go-ahead. In April 2019 the London Borough of Bexley was successful in persuading the Housing Secretary to make £4.85 million available for a feasibility study into options for delivering transport enhancements from Abbey Wood to Ebbsfleet.

---

[17] Crossrail Information Paper C5 Additional Safeguarding

Two consultation rounds have taken place with three options put forward for consideration: a low cost bus rapid transit scheme connecting with the existing railway; a high cost plan to extend all 12 trains an hour currently terminating at Abbey Wood on to Dartford by four-tracking the railway to Dartford; and an in-between plan that would see four of the 12 trains extended to Northfleet (with an enhanced pedestrian link to Ebbsfleet station). A further four trains would run to Gravesend with Elizabeth line services sharing track with North Kent services.

Under this proposal some existing national rail services would see their frequencies reduced or be withdrawn; the main cut would be between Dartford and Northfleet where there would be a two trains per hour reduction in the national rail service - offset by eight trains an hour running on an extended Elizabeth line. However, east of Northfleet, to Gravesend, the four trains an hour Elizabeth line service would be a direct replacement of the existing four national rail trains per hour.

Although this option requires compromises the opportunity to place communities on the Elizabeth line map, and offer them direct services into Docklands, central London and the west, appears to have persuaded many civic leaders that the benefits are superior and that this is a project that can be delivered. In October 2021 the leaders of Bexley, Dartford, and Gravesham councils, along with Kent County Council, jointly signed a letter that was submitted to the Department for Transport, along with an outline business case, for an option that would see eight of the current 12 trains per hour extended to Northfleet with four of the eight going on to Gravesend.[18] Post-pandemic rail industry finances may reduce the likelihood of government releasing funds for new projects but, given the level of support, pressure to deliver a south eastern extension of Crossrail is likely to persist. Picture the Elizabeth line of the future and chances are, at some point, trains will run beyond Abbey Wood.

---

[18] Ian Visits, Local councils opt for cheaper Crossrail extension into Kent, 3 November 2021

*Previous page:*

Old Oak Common station under construction for High Speed 2. Elizabeth line trains (seen stabled at Old Oak Common depot top left) will serve the new station and transport HS2 passengers onwards to central London.

*Picture courtesy of HS2 Ltd.*


*Facing page:*

*Station diagram courtesy of TfL.*

# Stations on the Elizabeth line

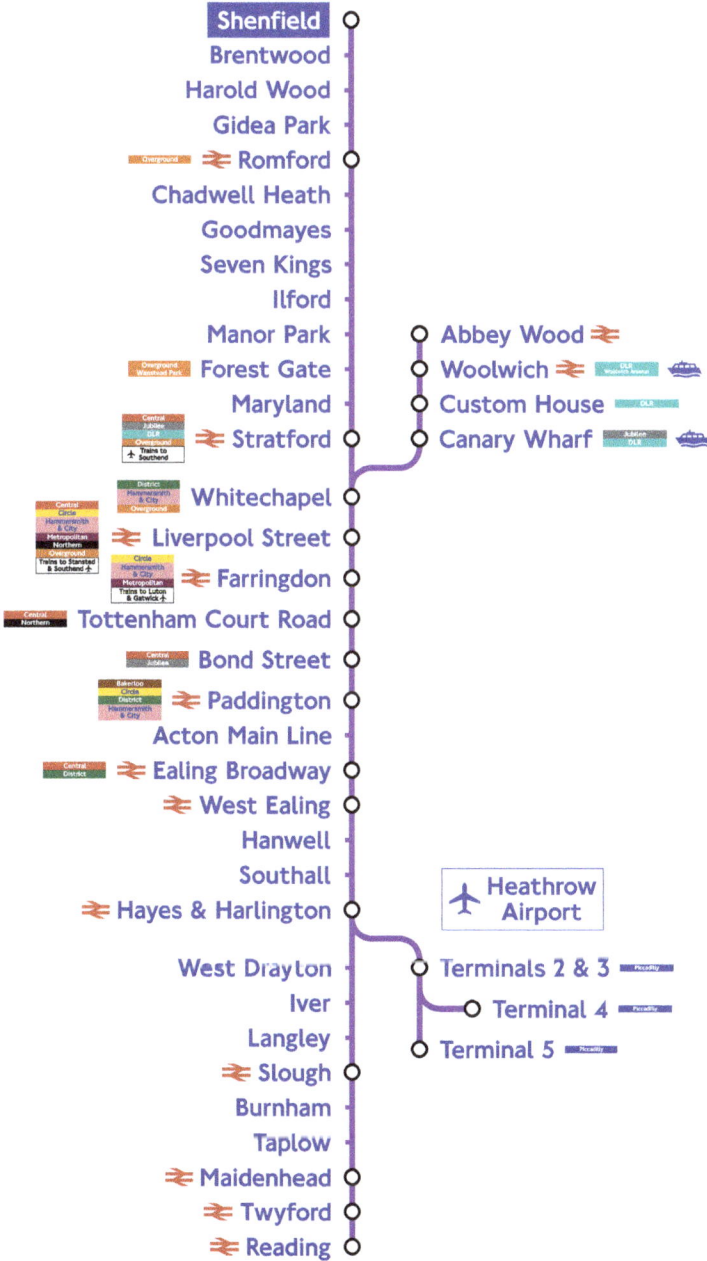

Shenfield

Brentwood

Harold Wood

Gidea Park

Overground ⇌ Romford

Chadwell Heath

Goodmayes

Seven Kings

Ilford

Manor Park      Abbey Wood ⇌

Overground Weavest Park | Forest Gate      Woolwich ⇌ DLR Woolwich Arsenal 🚢

Maryland      Custom House DLR

Central Jubilee DLR Overground ✈ Trains to Southend | ⇌ Stratford      Canary Wharf Jubilee DLR 🚢

District Hammersmith & City Overground | Whitechapel

Central Circle Hammersmith & City Metropolitan Northern Overground Trains to Stansted & Southend ✈ | ⇌ Liverpool Street

Circle Hammersmith & City Metropolitan Trains to Luton & Gatwick ✈ | ⇌ Farringdon

Central Northern | Tottenham Court Road

Central Jubilee | Bond Street

Bakerloo Circle District Hammersmith & City | ⇌ Paddington

Acton Main Line

Central District | ⇌ Ealing Broadway

⇌ West Ealing

Hanwell

Southall

✈ Heathrow Airport

⇌ Hayes & Harlington

West Drayton      Terminals 2 & 3 Piccadilly

Iver      Terminal 4 Piccadilly

Langley      Terminal 5 Piccadilly

⇌ Slough

Burnham

Taplow

⇌ Maidenhead

⇌ Twyford

⇌ Reading

www.ingramcontent.com/pod-product-compliance
Lightning Source LLC
Chambersburg PA
CBHW040411110426
42812CB00012B/2520